FOOD TEXTURE

First Edition

First printing, 1962
Second printing, 1999

by

Samuel A. Matz

December 1999

FOOD TEXTURE

First edition, Second printing

Copyright 1990 by SAMUEL A. MATZ

ISBN 0-942849-23-X

PAN-TECH INTERNATIONAL, INC.
P. O. Box 4548
McAllen, TX 78502

Printed in the United States of America

Contents

List of Illustrations

The Meaning and Measurement of Texture

CHAPTER 1

The Meaning and Importance of Texture in Foods

DEFINITION OF TEXTURE

The word "texture" is derived from a Latin root meaning "to weave" and it was evidently first applied in the English language to certain tactile and visual characteristics of fabrics. By analogy it came to have a more general meaning and was applied to other classes of objects including foods. Its exact connotation when applied to some of these substances is frequently unclear and nowhere is the lack of clarity more evident than in its application to foods.

According to the Taste Testing and Consumer Preference Committee of the Institute of Food Technologists (Kramer 1959), texture, as the term is used by persons engaged in sensory evaluations of foods and beverages, is the totality of "those properties of a foodstuff apprehended by the eyes and by the skin and muscle senses in the mouth, including the roughness, smoothness, graininess, etc." Unless this definition is considered to be implicitly limited by the examples which were quoted, it is extremely broad, encompassing not only the qualities commonly described as taste but also the totality of factors involved in appearance, including size, shape, and color. Even if taste is excluded as not being apprehended by one of the "skin or muscle senses in the mouth" (and this would be a doubtful interpretation of the latter clause), the sensations of temperature and pain would still have to be included as components of texture.

As a matter of fact, many workers in the food field seem to have taken a more restricted view of texture, adopting a meaning corresponding more nearly to the committee's definition of "mouthfeel," i.e.,

The mingled experience deriving from the sensations of the skin in the mouth after ingestion of a food or beverage. It relates to density, viscosity, surface tension, and other physical properties of the material being sampled.

From this point of view, texture would be considered to be part of, or similar to, the complex of properties which has been called kinesthetics. Kramer (1955) divided the quality factors of foods into three categories: (1) appearance factors, (2) kinesthetic factors, and (3) flavor factors. Earlier, Kramer (1951) had listed chewiness, fibrousness, succulence, and grittiness as the kinesthetic factors which are important to the objective testing of vegetable quality.

Other authors have attempted to define texture as the word is applied to foods. Ball *et al.* (1957) recognized that texture of meats had never been well defined and set about to provide adequate definitions for the term. According to these authors, a definition should meet the following requirements if it is to be of value:

(1) The definition should describe an attribute that is different from any attribute which figures in the definition of another quality characteristic, with which texture is associated, such as tenderness or juiciness.

(2) The definition should be consistent with the dictionary definition of the term.

(3) The attribute described should be important in the use to which it is applied.

(4) Examination results, based on the definition, should show a relationship to other attributes which, by the nature of their definitions, have a logical relationship to texture.

Based on these requirements, they developed two tentative definitions of texture, one based on appearance and the other based on "feel." The appearance definition, as applied to meat, was given as

texture of meat is the macroscopic appearance of meat tissue from the standpoint of smoothness or fineness of the grain. The "grain" is defined as the macroscopic appearance of cut surfaces of lean parts, which it is possible to describe as smooth, fine, rough, or coarse. Fineness is assumed to be dependent upon the size of the fiber bundles; the smaller the bundle under macroscopic examination, the finer the texture.

The "feel" definition was worded as follows:

Texture of cooked meat is the feel of smoothness or fineness of the muscle tissue in the mouth.

Scott Blair (1960), who has published extensively in the field of rheological properties of foods, made the following statements:

In certain industries, such as dairying, the term "texture" has a somewhat restricted definition. For the purposes of the present discussion, "texture" will be used in its widest denotation, to include "body," "consistency," etc.

In an earlier publication, Baron and Scott Blair (1953) used "body" and texture interchangeably in referring to a certain complex of physical characteristics in cheese and curd.

There is some disagreement among the practitioners in the various subdivisions of food technology as to the meaning of texture. The quotations from Ball et al. (1957), can be assumed to be representative of usage in the meats field, and Scott Blair doubtless defines the term correctly as it is used by dairy technologists, but the two definitions are not very much alike. In evaluating bakery products, texture is sometimes determined solely by tactile methods while "grain" is used to signify the visual roughness or smoothness (as well as some other qualities) (Dalby and Hill 1960) and "mastication" is sometimes used as approximately equal to "mouthfeel" as the latter term was defined by the Institute of Food Technologists. However, many experimental bakers prefer to use "texture" in a broader sense by letting it apply to mouthfeel as well as to the characteristics determined by touch.

Although the evidence is admittedly rather meager, it appears to the author that most members of consumer panels and of the general public which he has contacted regard food texture as a quality determinable principally by sensations in the mouth, particularly by resistance to mastication. The preponderance of current usage in technological publications in the field seems to favor a meaning of the word "texture" which corresponds approximately to the Institute of Food Technologists Committee definition of mouthfeel.

The term texture, as used by the author in the following pages, should be understood by the reader to mean those perceptions which constitute the evaluation of a food's *physical* characteristics by the skin or muscle senses of the buccal cavity, excepting the sensations of temperature or pain. This definition is intended to exclude the chemically initiated sensations of taste and, of course, does not regard texture as having a visual component.

SENSES RESPONSIBLE FOR TEXTURE PERCEPTION

Oldfield (1960) divides into three groups the sense organs responsible for perceiving texture:

(1) those in the superficial structures of the mouth (the hard and soft palate, tongue, and gums), (2) those around the roots of the teeth (in the peridontal membrane), and (3) those in the muscles and tendons used in mastication.

The superficial structures respond to the smoothness, roughness, stickiness, slickness, and related surface characteristics of the ingested foods while the receptors in the peridontal membrane and in the muscles and tendons are affected more by the mass behavior of the substance, that is, its relative elasticity, brittleness, etc. In addition, the tongue probably plays a considerable role in sensing the viscosity of fluids. It is beyond the scope of this volume to go into details of the anatomical and physiological bases of texture judgments.

IMPORTANCE OF TEXTURE

A consumer's reaction to a food, apart from extraneous influences such as advertising and custom, would seem to be governed entirely by the sets of qualities which are summed up in the words texture, flavor, and appearance. To most consumers, appearance is the most important of these factors because the initial awareness of the substance is usually the result of visual perception, and this initial impression probably influences subsequent judgments regarding the other factors. It is difficult to evaluate the relative importance of flavor and texture. Indeed, such a question may not even be meaningful, since it is impossible to compare the two sets of qualities on equivalent scales. However, it is meaningful to ask which of the two sets of qualities has more influence on acceptance when they are varied within the ranges expected in normal commercial production.

There have been very few publications which described attempts to determine the relative contribution of flavor and texture to the overall acceptability of foods. Campbell (1956) found that the desirable characteristic of beef most important to consumers was tenderness, while tastiness was second, and juiciness (another texture-related quality) was third. The most influential undesirable characteristics were toughness, dryness, and lack of flavor, in that order. Rhodes *et al.* (1955) concluded that "tenderness was the outstandingly important eating characteristic desired in beef," based on the results of consumer preference studies. Several other workers have reached the same conclusions with regard to meat, but the situation is not as clear when other foods are being considered.

It is probable that the range of fluctuations in texture encountered in commercial foods is less important in influencing acceptability than is the usual variation in appearance. That is, from a given number of samples randomly selected, more will be rejected by the average consumer for appearance defects than for texture deficiences. On the other hand, an acceptable flavor quality may be easier to achieve than a satisfactory texture in many foods. For example, the average consumer seems to be rather tolerant of changes in bread flavor, but relatively small (as measured objectively) changes in texture can cause rejection. Far more bread samples are discarded because of texture deterioration than for flavor flaws.

TEXTURE AND CULTURE

The foods consumed in greatest quantity by nearly every cultural group throughout the world tend to be white in color, mild in flavor, and soft in texture. Rice, alimentary pastes, potatoes, oatmeal and other gruels, hominy grits, poi, and milk represent the closest approaches to the ideal which economic and agricultural limitations have allowed various groups to make. Commercial white bread is another outstanding example of this universal trend. Most cultural groups which consume yeast-leavened breads appear to favor the whitest, softest, and mildest-flavored product which their technology can produce and their standard of living can afford. It is true that there is some lag in this trend where tradition—always a very strong force in directing food selection—inclines the consumer toward the acceptance of a form of bread derived from a more primitive technology. Furthermore, in prosperous societies many unusual varieties of food will find a certain demand based on their novelty and their appeal to jaded tastes. However, the voluntary consumption of small percentages of specialty breads deviating in one or more respects from the "best" commercial product certainly does not vitiate the argument that most people prefer their bread soft, very white, and bland in flavor. There will always be consumers who feel that the unusual is bound to be better, especially if it is also more expensive, but there is no more reason why a denser, chewier, stronger-flavored bread should be considered "better" than the ordinary commercial product than there is for preferring a tough steak over a tender one.

Although the foods consumed in greatest quantity tend to be white, soft, and insipid, the supplementary foods may vary widely in the characteristic of flavor, appearance, and texture. Although the Oriental may wish for his rice to be as white, as flavorless, as tender, and as plentiful as possible, he would also like to be able to supplement it with relatively small amounts of sauces, pickled vegetables, fish, pork, or poultry having distinctive and often strong flavors and crisp, crunchy, or chewy textures. These supplementary foods always tend to be much more frequently changed (from meal to meal and from day to day) and to be consumed in much smaller quantities than the basic foods.

Where supplemental foods are consumed in considerable amounts, and, in some cases, take the place of basic foods, as in the United States, the same trend to pale colors, bland flavors, and tender textures is often observed. Thus we, as a people consume much more cottage cheese than we do aged Cheddar or Roquefort. A few other examples may be of interest. In poultry, the trend has long been to fowl with more of the less flavorful white meat, and, of course, with tenderer flesh. In hams, the trend has been for a long time away from the drier, stringier, highly-flavored varieties to the presently popular type which in some cases resembles a pink-colored jelly having a mildly salty flavor. In the selection of fruits and vegetables, the tendency is certainly less pronounced so far as color is concerned, even though it becomes evident in such foods as celery. The trend of flavor preferences in fruits and vegetable consumption is also hard to pin down, but an increasing demand for softer texture is generally apparent and is frequently translated into economic terms, i.e., the softer types of a given food are rated as more valuable than the harsher textured examples. Of course, the changes in preference are gradual, and there always remains a limit beyond which greater softness becomes a liability. These limits are usually established by customs and familiarity—what the consumer has been led to expect from the food in the past.

CLASSIFICATION OF TEXTURAL QUALITIES

In order to permit the discussion of texture to be presented in an orderly and concise manner, it was necessary to devise a classification scheme which was based on sound logical and practical considerations. The literature was not of much help in this regard. The very few collections of papers on the subject of food texture which are available have been put together in a rather desultory fashion. For example, "Texture in Foods" (Anon. 1960), contains the following chapters (or papers): (1) Perception in the mouth, (2) Foods of simple structure (sugar syrups, boiled sweets), (3) Emulsions and related dispersions, (4) Foams in confectionery with special reference to marshmallows, (5) Gels— with special reference to pectin gels, (6) Texture in bread and flour confectionery, (7) Chocolate, (8) Scientific principles in relation to instrumental measurement of textural properties, (9) Measurement of meat tenderness, (10) Texture change in fish and its measurement, (11) Effect of chemical constitution on texture of peas, (12) Texture in cooked potatoes, (13) Some factors involved in the texture of plant tissues, and (14) Strength in the egg.

An earlier book on the rheology of foodstuffs (Scott Blair 1953) contained the following contributions: (1) Starch, (2) Cereals, (3) Rheology of milk, cream, ice cream mixes, and similar products, (4) The consistency of butter, (5) Rheology of cheese, (6) The rheology of honey, (7) The rheology of certain miscellaneous food products, and (8) Psycho-rheology of foodstuffs. Although these volumes are quite valuable in many respects, they offer little assistance in the formulation of a logical classification for food textural qualities.

Since there was no known precedent for the desired classification scheme, it was necessary to develop one based on fundamental principles. Several possible schemes were considered before a satisfactory one was developed. These are discussed in the following numbered paragraphs.

(1) Classification according to the type of organism from which the food was derived: All foods obtained from fish would be included in one category while all those obtained from fruits would be in another, etc. The only advantage of such a classification method is its simplicity of application. It would unquestionably be easy to decide upon the category into which each food would be fitted, but the diversity of textural types

found in foods taken from the same group of organisms means that the categories would be too diffuse to be meaningful.

(2) Classification according to the type of tissue constituting the food: Products consisting of muscle tissue would be placed in one group, those consisting mostly of adipose tissue would be placed in another group, etc. Objections to this system arise from its frequent inability to permit predictions of the type of sensory reaction on the basis of the category in which a food is placed. That is, there is no dependable relationship between the kind of texture exhibited by a food and the tissue from which the food was obtained.

(3) Classification according to the chemical composition of the food: According to this method, foods would be classified according to their principal constituent, e.g., proteins, starches, etc. The main problem with this system is that it allows too much overlap between classes. Many foods have large proportions of several categories of compounds, all of which play some role in determining texture. In addition, the constituents having the most influence on texture may be present in relatively minor amounts.

(4) Classification according to the physical structure of the food: Fibrous foods might constitute one class, while gels would be another logical group, etc. If the categories were carefully chosen, a system of this type should perform reasonably well in separating foods of fundamentally different textural quality. Some overlapping of categories is possible since many foods consist of mixed structural elements, but the predominating structure should be easy to select in most cases.

(5) Classification according to sensory dimensions: Since texture is an organoleptic[1] characteristic, it would seem to be preferable to classify the various aspects of it on a sensory basis. However, the dimensions of texture are even less well known than those of odor or taste. There is not enough basic information on the subject to permit a meaningful classification in this manner. Nevertheless, it should be possible to group foods according to subjective evaluations of their textural similarities even though the areas in which these similarities occur are poorly understood.

The decision finally reached was to combine the structural with the sensory approach in the development of categories which seem to be fairly clearly demarcated on a sensory basis (i.e., most experts would relate the category description to the same foods with a minimum amount of indecision) and also have a common macroscopic structure. The categories finally chosen are not expected to meet with universal approval

---

[1] The author is aware of the objections raised by some psychologists to the use of this word in the sense it is used here, but it seems to me to fill a semantic need.

or persist entirely unchanged after criticism by experts in the field. However, it is believed that it is essential to make a first approach to a systematic classification of food textural qualities so as to bring some order to this presently rather chaotic field.

The categories of food textural qualities used in this volume are:

(1) Liquids—viscosity being the principal textural feature of this class.

(2) Gels—elasticity being a prominent feature of the rheological properties of these foods.

(3) Fibrous foods—macroscopic fibers being perceptible individually or in the mass.

(4) Agglomerates of turgid cells—in which whole cell properties exert the major influence.

(5) Unctuous foods—fatty substances and other foods resembling them in textural characteristics.

(6) Friable structures—those which readily break down to small irregular particles upon mastication.

(7) Glassy foods—hard, homogeneous masses which are often consumed by slowly dissolving them in the mouth.

(8) Agglomerates of gas-filled vesicles—food foams and sponges.

(9) Combinations of the preceding structures—no one structural element predominating.

BIBLIOGRAPHY

ANDERSON, E. E. 1958. Scoring and ranking. *In* Flavor Research and Food Acceptance. Reinhold Publishing Corp., New York.

ANON. 1960. Texture in Foods. Soc. Chem. Ind. Monograph 7.

BALL, C. O., CLAUSS, W. E., and STIER, E. F. 1957. Factors affecting quality of prepackaged meats. I. Physical and organoleptic tests. Food Technol. *11*, 277–283.

BARON, M., and SCOTT BLAIR, G. W. 1953. Rheology of cheese and curd. *In* Foodstuffs: Their Plasticity, Fluidity, and Consistency. Edited by G. W. Scott Blair. Interscience Publishers, New York.

CAMPBELL, G. W. 1956. Consumer acceptance of beef. A controlled retail store experiment. Ariz. Univ. Agr. Expt. Sta. Rept. *145.*

DALBY, G., and HILL, G. 1960. Quality testing of bakery products. *In* Bakery Technology and Engineering. Edited by S. A. Matz. Avi Publishing Co., Westport, Conn.

EHRENBERG, A. S. C. 1955. Descriptive terms and grading systems. Food Research *20*, 298–300.

KRAMER, A. 1951. Objective testing of vegetable quality. Food Technol. *5*, 265–269.

KRAMER, A. 1955. Food quality and quality control. *In* Handbook of Food and Agriculture. Reinhold Publishing Corp., New York.

KRAMER, A. 1959. Glossary of some terms used in the sensory (panel) evaluation of foods and beverages. Food Technol. *13*, 733–736.

KROPF, D. H., and GRAF, R. L. 1959. Interrelationships of subjective, chemical, and sensory evaluations of beef quality. Food Technol. *13*, 492–495.

OLDFIELD, R. C. 1960. Perception in the mouth. *In* Texture in Foods. Soc. Chem. Ind. Monograph 7.

RHODES, V. J., KIEHL, E. R., and BRODY, D. E. 1955. Visual preferences for grades of retail beef cuts. Missouri Univ. Agr. Expt. Sta. Research Bull. *583*.

SCOTT BLAIR, G. W. 1953. Foodstuffs: Their Plasticity, Fluidity, and Consistency. Interscience Publishers, New York.

SCOTT BLAIR, G. W. 1960. Scientific principles in relation to instrumental measurement of textural properties. Soc. Chem. Ind. Monograph 7.

SPERRING, D. D., PLATT, W. T., and HINER, R. L. 1956. Tenderness in beef muscle as measured by pressure. Food Technol. *13*, 155–158.

TRESSLER, D. K., and JOSLYN, M. A. 1954. The Chemistry and Technology of Fruit and Vegetable Juice Production. Avi Publishing Co., Westport, Conn.

CHAPTER 2

Measuring Food Texture

INTRODUCTION

Texture has long been recognized as an important element in the total sensory impression obtained during the consumption of foods. In spite of this knowledge, texture requirements have not found their way into specifications and standards for food products as often as have limitations based on appearance and flavor. The difficulties involved in accurately assessing textural quality, either objectively or subjectively, have often provided the excuse for such omissions.

It is becoming increasingly more evident that some form of texture measurement is highly desirable in the grading of nearly all foods, and that the problems involved in setting up workable specifications cannot be allowed to hinder their development. Consequently, efforts to find practical testing procedures and to establish satisfactory quality limits have been greatly accelerated in the past few years.

The plan followed in this chapter is to discuss subjective and objective measurements in separate sections, in each case covering the theory and general considerations first and then describing the practical measuring techniques which have been developed. In some instances, it has seemed to be more worthwhile to include the discussion of some specific technique in a later chapter, especially when its application is mostly restricted to a particular kind of food.

SUBJECTIVE MEASUREMENTS OF TEXTURE

Sensory tests are probably the only way by which the investigator can get a meaningful evaluation of texture as a whole. In the final analysis, it is always the consumer's judgment which must be relied upon to decide whether the texture of a food is good, bad, or indifferent. No texture preference test, subjective or objective, has much value unless it yields results which bear some relationship to the rating which would be given to the sample by the "average" consumer.

Although it appears that sensory tests performed by relatively large panels are of more value in predicting consumer reactions, expert or single judgment evaluations of textural qualities can be useful for control purposes, and even, to some extent, for guiding product improvement studies, if their limitations are kept clearly in mind at all times. The results of expert or small panel tests should be checked as often as possible by comparing them with the results of consumer-type panel tests.

In the following section, the various kinds of tests and their value (according to the author's opinion) in the situations likely to be met in product development and research work are reviewed and discussed.

Organoleptic Testing With Large Panels

Much of the following discussion on sensory testing using large panels will refer to conditions encountered, and conclusions reached, in flavor tests. The bulk of available information on sensory testing lies in this area. However, a substantial part of the information would appear to be applicable to texture evaluation as well. Many investigators have included questions on texture in questionnaires used in panel tests of foods and beverages and have obtained results which seemed to be meaningful. Evidently the average respondent understands what is meant by texture in relation to these products, and is able to evaluate this characteristic, to his own satisfaction at least.

In most cases, the goal desired in setting up large consumer-type panels is to make them as representative of the general population as it is possible to do within the limitations imposed by time and economic considerations. Some bias will always be present in any test group in spite of all attempts to assure random selection of members. Even the characteristic of willingness-to-test which renders a person available for the panel sets him apart from the general population. If selection is made from a group of individuals employed in a certain industry, the potential members are still more non-representative. If such a panel is then further restricted by picking, e.g., only office personnel, the error is compounded. A plan frequently used by marketing research groups is to choose a community deemed to be fairly representative of the population in question and then to distribute samples to residences picked by some random selection scheme. In addition to its other drawbacks, in many cases this method is prohibitive from the economic standpoint.

Geographical prejudices and traditions can be very strong influences in biasing food preferences and yet these factors are often ignored in evaluating the results of panel tests. For example, it would be clearly unsuitable to test a cornbread mix destined for southern consumption by using a panel selected from residents of a large northern city. Testing preferences for frozen ocean fish using mid-western consumer panels would be another example of geographical error, since persons in this region are more familiar with (and may prefer) either fresh water fish, unfrozen, or canned ocean fish of a few limited types.

Age of the respondent has been shown by several investigators to be a factor affecting food evaluations. Sex of the tester has a lesser influence but may at times affect results significantly. Economic level, or its correlate education, have been shown to have effects. Race, religious taboos, and national origins are unquestionably related to food preferences although these differences are decreasing and will doubtless become much less important as better food distribution, more effective advertising, and consumer education bring food preferences to a common level throughout the country.

Sample Presentation Schemes and Rating Scales.—The simplest test scheme is the presentation of a single sample for evaluation as either acceptable or unacceptable. Unfortunately, this uncomplicated procedure often does not yield the desired information. Modifications having somewhat more flexibility have been proposed from time to time. Psychologists generally classify sensory tests as either affective or discriminative. The first category includes those tests in which the subject is asked to give some sort of a rating (selected from a hedonic scale, for example) to the sample, and the second class includes those tests in which a number of samples are compared and the different one(s) selected. The latter group has been reviewed in some detail by Peryam (1958) and the different types of tests will be discussed briefly in the following section in order to orient the reader in some of the terminology of this discipline. Limitations of space precluded the addition of "how-to" information. This should be sought in the standard works of the field, as given in the bibliography.

The paired difference test involves the presentation of two unknowns, either simultaneously or successively, in random order. The subject is directed to select one of the two according to a given criterion, e.g., to select the saltier sample. In the duo-trio difference test, two unknowns are also presented, but they are accompanied by a standard, or known, which is equivalent to one of the unknowns. The tester is directed to match the standard with one of the unknowns. In the dual standard test, both of the unknowns are represented by equivalent identified samples and the subject is supposed to match each of the unknowns with the proper standard.

The triangle difference test is a popular scheme using an approach which is slightly different from those described above. In these tests, the subject is confronted with three samples, two (only) of which are the same. The subject's task is to select the different sample.

Peryam also described the multiple standards method, single stimulus ("A—not A"), multiple pairs, and variations of these patterns. Any of these methods may have some special value for a particular situation, but, except for statistical efficiency, the results obtained can be secured just as well by another of the procedures described above.

When experimental samples differ from the controls in a number of different characteristics, or when the variations are quite extreme, difference tests are not often applicable since all, or nearly all, panel members will be able to detect the odd samples. In such cases, scoring or rating tests are used. These tests require the subject to indicate on a preestablished scale his opinion of the overall quality, or of the degree of excellence of single characteristics.

One of the difficult problems encountered in setting up a rating test is the design of a score sheet or scale which will include meaningful intervals of quality. It is obviously not suitable to ask the panel member to rate the sample on a scale running from 0 to 100 per cent, with the latter point indicating perfection. What is 50 per cent of perfection—does it represent a marginally acceptable product or is it a good commercial sample? In attempts to overcome such uncertainties, it has been the usual practice to assign descriptive adjectives to each point on the scale. One of the most popular rating plans which has been developed to date is the nine-point hedonic scale. The lowest point on this scale (given a value of one) is described as "Dislike extremely," while "Like extremely" is given a score of nine, and "Neither like nor dislike" is scored as five. Numerous studies, by Peryam and his co-workers, have proved the usefulness of this rating method for predicting preference of a wide variety of ration items. It is also very adaptable to statistical manipulation.

A preference method somewhat analogous to the paired difference method involves presenting an experimental sample and a control to the panel member and asking for a decision as to his preference, either overall or on the basis of some specific quality factor. Such a rating component can also be combined with the triangle test and most of the other difference tests by requiring the tester to state his answer to the question:

Do you prefer the odd sample or the paired samples?

However, according to Pilgrim (1961),

The difference-preference test has been shown to be biased and should not be used.

Ranking is a method of obtaining product evaluations which can frequently be more useful than the difference tests and rating tests described previously. It does not depend upon a rating scale having artificial intervals of uncertain equivalence. In principle, ranking methods involve presenting any number of samples at random, in sequence or simultaneously, to the tester who is directed to arrange the identification symbols in the order of his preference for the samples. From a practical standpoint, an important advantage of ranking is the ease with which a large number of samples can be tested simultaneously. Recent papers (Kramer 1956, 1960; Bradley and Kramer 1957) described statistical procedures for determining the significance of ranking data without the necessity for resorting to tedious calculations.

Gridgeman (1961) attempted to evaluate the relative efficiencies of a pair comparison test with "straight" preferences, a pair comparison test including degrees of preferences, and a rating scheme in detecting differences in flavor. The data of Murphy *et al.* (1957) were used. The results reproduced below were obtained:

	Relative Efficiencies	
	Egg Tests	Fruit Tests
Simple pair preferences	100	100
Pair preferences with degrees	236	142
Rating	0	44

Gridgeman commented

In all germane sensory work there is a "natural" scale unit: it is the residual standard deviation of observations taken on the prescribed interval scale—whatever that may be, rank numbering, graded categories, or degree of preference . . . . One consequence of this natural unit is that F values (variance ratios) offer a rational device for the assessment of efficiency.

The Testing Environment.—If the samples are to be tested in the subject's residence, as in home placement studies, the environment must be regarded as essentially uncontrollable even though directions to test only at some specific time of day or in a certain manner are given. When the test is to be conducted in a facility set aside for that purpose, with isolation booths, controlled lights, pass-through service, and all of the paraphernalia that usually accompany them, the conditions can be rigidly standardized with resultant improvement in reproducibility and significance. Of course, such arrangements make no provision for estimating the effects of variations in preparation techniques (cooking times and temperatures, etc.) which would inevitably accompany use in the home of foods other than ready-to-eat products.

It seems fairly obvious that the time of day when the samples are presented will influence preference judgments, particularly if the product is strongly associated with one of the normal meals. Furthermore, testing immediately after the subjects have consumed a meal would seem likely to yield lower preference ratings than tests of the same product just before mealtime. Ranking and difference tests would presumably not be as subject to these influences. However, Pilgrim (1961) says,

Ranking is subject to the same influences but the effect is obscured by the method. The effect can be, and normally is, removed from the ratings in the analysis.

If flavor (or texture) is the only sensory property of interest, means exist for readily counteracting the effect of differing colors on consumer judgments. In these cases, the tests can be conducted under colored lights of low intensity. It is sometimes possible to color the samples artificially to a neutral hue by adding combinations of food dyes. Cancelling out differences in texture which might bias flavor judgments is apt to be considerably more difficult. Homogenization might be tried but Kramer *et al.* (1961) showed, in the case of peaches, that the ability of consumers to detect flavor differences correctly decreased as the size of the pieces decreased (sliced, diced, and puréed).

Temperature of the samples can affect acceptability, and coffee samples (for example) which vary only slightly in temperature may be evaluated on that basis alone by some of the testers, while the temperature of ice cream samples not only has a direct effect on preference but also an indirect influence based on the alteration in texture which accompanies the temperature changes. Gridgeman (1958) showed there was a general tendency for subjects to confuse temperature differences with taste differences. In his experiments, there was a tendency for both hot and cold—but particularly the latter—to be associated with a decrease of stimulus. In other words, both hot and cold reduced sensitivity. The subjects with the highest taste acuity tended to exhibit the least bias on a temperature basis.

Gridgeman also showed that a positional bias existed. This may be right or left hand, depending on the subject, and may vary. Remedial measures include balancing the two (or more) possible presentation arrangements, and randomizing within that restriction.

There has been some question as to whether or not the ability of a panel member to discriminate between samples having only small differences in acceptability tends to decrease as the number of samples is increased. Maple syrup is said to be a product which causes a rapid drop in sensitivity after a few ingestions. On the other hand, Sather and Calvin (1960) showed that, for mild products such as green beans, hamburger, and peaches, up to 20 samples could be tasted in one session with no decrease in the judges' ability to discriminate between the samples. Kramer *et al.* (1961) recommended presentation of up to nine samples per sitting to increase testing efficiency. These authors add that increased precision is attained only by increasing the number of panelists, rather than by replication, when consumer panels are used, in contradistinction to the situation existing when expert panels are used.

The ability of a given subject to evaluate flavor or texture differences may vary with time. Baker *et al.* (1960) stated

". . . the probability of distinguishing A from B for any one person or group of persons increases from zero to one over a rather wide range for the difference between A and B. There is also the possibility of change in preference as the testing proceeds. This latter situation could be guarded against only by rather elaborate techniques . . . . One method of proceeding is to test each of the *n* consumers of the random sample once and record the stated preference. Consumers should be forced to state a preference because of subliminal considerations. Then test each again. Those who reverse their preferences are temporarily sorted out. Similar steps are completed with reversals sorted out until five consistent statements are made concerning the preference for A or B. Such people are likely to have a real preference for A or a real preference for B as the case may be. If very few prefer A or very few prefer B we conclude that the preference is essentially in one direction and this is a very valuable thing to know. If the number preferring A or B are both small, we can conclude there is no difference in preference . . . . In choosing five consistent judgments we are operating at about the three per cent level.

If one change in preference is permitted in the subjects' series, the three per cent level of confidence can be attained with eight tests.

"Expert" and Other Special Panels

Attempts to simplify and "improve" consumer panels have led to the use of expert panels composed of persons with long experience or special training in detecting and identifying small differences in the organoleptic characteristics of certain products. The best known examples, such as coffee blenders, wine tasters, and artificial flavor compounding artists, are not much concerned with texture. Although these highly skilled individuals or groups can greatly expedite the testing of samples, they should not in any case be regarded as superior to, or more sensitive than, consumer panels which have been properly assembled and intelligently employed, and whose results have been suitably analyzed. The prejudices, biases, and lack of sensitivity to certain flavor nuances of representative consumer panels are important and necessary parts of their predictions of "average consumer" reactions. Experts may lack some or all of these characteristics and may also possess an inherent bias which introduces systematic errors into their judgments of flavor or texture quality. It is for this reason that the results obtained from expert panels need a more penetrating and cautious interpretation than those obtained from properly constituted consumer panels. In the latter case, much of the weighting has already been done. Where an expert panel deviates from the judgments of a consumer panel, its data are, in fact, less valid from a marketing standpoint, even though it can be shown that the expert panel can detect smaller increments of some quality which can be verified by objective tests.

Members of expert panels must be adequately screened and trained if they are to function satisfactorily. As stated by Kramer *et al.* (1961) expert panelists cannot be obtained after one screening, but need training and experience in repeated screenings before they achieve expertness in detecting flavor differences. They also say that a small number of panelists are adequate for detecting flavor differences since the degree of precision which is required may be secured by replicated testings of the same sample.

Even less useful than expert panels are those testing groups composed of persons whose qualification is principally their rank in the company. Frequently, such panels will be constituted mostly of lawyers, accountants, and sales personnel. In general, executives with such backgrounds are incapable of making, or even of understanding the desirability of, unprejudiced judgments of proposed new products. They are liable to be guided by an overpowering set of ingrained prejudices, or by policies and traditions, which bear little or no relationship to the likes and dislikes of the average consumer. Furthermore, their living conditions and dietary habits are likely to be quite atypical, and to influence their judgments on food quality. Some of the problems encountered when product development groups are guided by panels of this sort have been well described by Hower (1958). Most progressive companies now recognize the necessity for putting product evaluation under the control of persons who will not be directly responsible for selling, advertising, manufacturing, or financing the product and who can take full advantage of the available statistical tools.

Does the expert panel have any place in the testing program of a food development team? The answer is a qualified "yes." For rapid day-to-day screening of large numbers of preliminary samples, an expert panel may be necessary because of the practical difficulties of rapidly testing by the consumer-type panel of as many samples as desired. Leaving the screening entirely to the laboratory personnel who are directly engaged in the development work may be hazardous because of their often apparent tendency to develop enthusiasms which are not based entirely on the sensory properties of the food. Furthermore, an expert panel can, in the most efficacious examples of the type, provide immediate guidance as to the *general* direction in which changes should be made in order to secure greater consumer acceptance.

The minimum training for an expert should include an orientation discussion in which the goals of the product developers are explained, possible criteria for judging are suggested, and the test plan outlined, and a series of familiarization sessions and comparison samples. "Experts" can be chosen from among those people who have had close technical contact with similar products for some years, or they can be selected on the basis of aptitude and sensitivity by running a series of elimination tests in which samples differing in progressively smaller increments of a pertinent factor are offered for evaluation.

It would seem that expert panels would be less subject to satiation and to the influence of position of sample presentation as well as other non-pertinent factors. If properly selected, they should also be able to detect smaller changes and to detect changes of the same magnitude more consistently than consumer panels of about the same size. From a practical point of view, they are usually easier to assemble at a given time than a larger consumer panel, and the total time "lost" in testing is often less. It is probably advisable in expert testing, as in large-panel testing, to isolate the subject while the tests are being made. Allowing a free flow of conversation and the discussion of results tends to distract the taster and, in addition, frequently affects his decision in the direction of that formed by the dominant personality in the group.

The expert panel can be of particular value in the assessment of product changes which are important from a theoretical standpoint but for which there is available no adequate instrumentation. In other words, the panel can function as a piece of testing equipment. In such situations the expert panel has more utility than the consumer panel because of its directed sensitivity. One method of testing for special characteristics by expert panels which has been developed in recent years to a high degree of precision is the so-called flavor profile method (reviewed by Caul *et al.* 1958). In essence, this method of testing involves a semi-quantitative analysis, by a trained panel, of the components or "notes" making up the flavor or aroma of a product. It is probably most valuable when applied to breaking down the formulas of competitive products and in identifying the source of adventitious off-flavors in one's own product. It would seem to have less utility for straightforward developmental work with foods (as opposed, say, to compounded flavors). Panels using this procedure depend upon oral review and evaluation of findings at the time of testing. Caul *et al.* explain that this procedure has, in practice, led to superior results in spite of the apparent difficulties introduced by the oral consultation. The profile method has apparently not been applied to the evaluation of texture, although it probably could be adapted to the examination of these characteristics.

The ultimate extrapolation of the expert panel is the single authority such as the tea taster (and like experts). These persons can develop truly remarkable sensitivity over the years and are often invaluable in maintaining unchanged the flavor characteristics of a high quality blend. However, it should never be assumed that their standards of taste (or texture) are identical with those of the general public. Their constant preoccupation with the most subtle details of flavor and the extreme acuteness of their perception can create a wide gulf between their judgments of quality and acceptability, and the preferences of the general public.

OBJECTIVE MEASUREMENTS OF TEXTURE

Physical Properties Related to Texture

Among the physical properties clearly related to texture are viscosity, elasticity, particle size, and surface qualities such as roughness and stickiness. Density, temperature, and some other properties may have an indirect influence on subjective evaluations of texture. The rheological characteristics of viscosity and elasticity often have been found to be closely related to panel test results, especially when foods of rather simple structure are being investigated.

Viscosity is usually defined as the resistance encountered by a layer of fluid as it moves over another layer of the fluid. It is the internal friction of the liquid. The equation, $f = \eta (Au/s)$, has been empirically derived to explain some of the observed viscosity relationships. This equation indicates that the tangential force, f, required to maintain a constant difference between the velocities of two parallel layers of a liquid moving in the same direction varies directly with the difference in velocity, u, and the area, A, of the surface of contact of the two layers, and varies inversely as the distance, s, between the two liquid layers. The symbol η is a proportionality factor known as the coefficient of viscosity. Its reciprocal, $\phi = 1/\eta$, is called fluidity.

Foods (and other substances) which have rheological properties that can be described by the simple expression given above are called Newtonian fluids. Substances are categorized as Newtonian or non-Newtonian depending upon their behavior under imposed shearing forces. Consider the case where two parallel plates, each of area A, are separated by a distance R, the intervening space being filled by the liquid in question. If a force F is applied to one of the plates, a *shearing stress* F/A (usually designated by τ) will result. This stress moves the plate with a velocity, V, which is related to the distance between the plates, so that the shearing rate, or *rate of shear*, can be described as $-dV/dR$. If the rate of shear is related by the same constant to the shear stress at all values of the latter (provided that laminar flow limits are not exceeded), that is $F/A = \eta (-dV/dR)$, then the fluid is said to be Newtonian. This is the same as saying that the coefficient of viscosity of Newtonian fluids remains the same at all shearing stress values (until turbulent flow develops). One consequence of this relationship is that a plot of τ vs. $-dV/dR$ yields a straight line passing through the origin. Non-Newtonian substances do not yield such a plot. Newtonian fluids include most simple solutions such as sucrose syrups, ethanol solutions, broths and bouillons, carbonated beverages, and skim milk. Most foods are non-Newtonian.

Metzner (1956) gives the following classification of non-Newtonian substances: (A) Time-independent non-Newtonians: (1) Bingham plastics or plastic fluids, (2) pseudoplastic fluids, (3) dilatant fluids. (B) Time-dependent non-Newtonians: (1) rheopectic fluids, (2) thixotropic fluids.

Fig. 1 illustrates some of these relationships (diagram adapted from Daugherty 1961).

Much of the following discussion has been derived from the treatments of Metzner (1956, 1961), Charm (1960), and Bingham (1922).

Bingham plastics require the application of a definite shearing force before any noticeable flow takes place, that is, $\tau = b(-dV/dR) + C$, where C is the stress which must be exceeded before flow occurs and b is a proportionality factor substituted for the inapplicable viscosity coefficient. After the yield stress has been applied, the relation of F/A to $-dV/dR$ is linear. A plot of shear stress against rate of shear for a hypothetical Bingham plastic is shown in Fig. 1. On a molecular level, Bingham plastics are envisioned as consisting of three-dimensional networks when they are at rest. Applied forces are resisted up to the point at which the network breaks down; the flow then becomes essentially Newtonian. Metzner (1961) stated "while Bingham plastic behavior is important historically, no convincing evidence is available to support the suggestion that any real fluids [aré Bingham plastics]."

Pseudoplastic materials (and this category includes most non-Newtonian foods) are distinguished by shearing rates which never become linear at any imposed shear stress. Instead, the shearing rate tends to increase faster than the shearing stress. A mathematical expression of this relationship is: $\tau = b(-dV/dR)^s$, where s, the so-called pseudoplasticity constant, is less than 1 but more than 0. A factor called differential viscosity, which has been found to have some value for practical purposes, can be derived from the preceding equation. It is $\eta_d = d\tau/d(dV/dR)$. The probable explanation for the rheological behavior of pseudoplastic fluids, according to Metzner, is that particles or molecules which are initially randomly oriented became increasingly better aligned as shear is applied. Their interactions and the contributions of these interactions to the apparent viscosity of the fluid accordingly became less with increasing shearing stress.

Dilatant fluids exhibit behavior which is the opposite of that of pseudo-plastics. That is, their apparent viscosity increases with increasing shear stress. Mathematically expressed, they behave in accordance with the equation $\tau = b(-dV/dR)^s$ where s is greater than 1 or less than 0. The

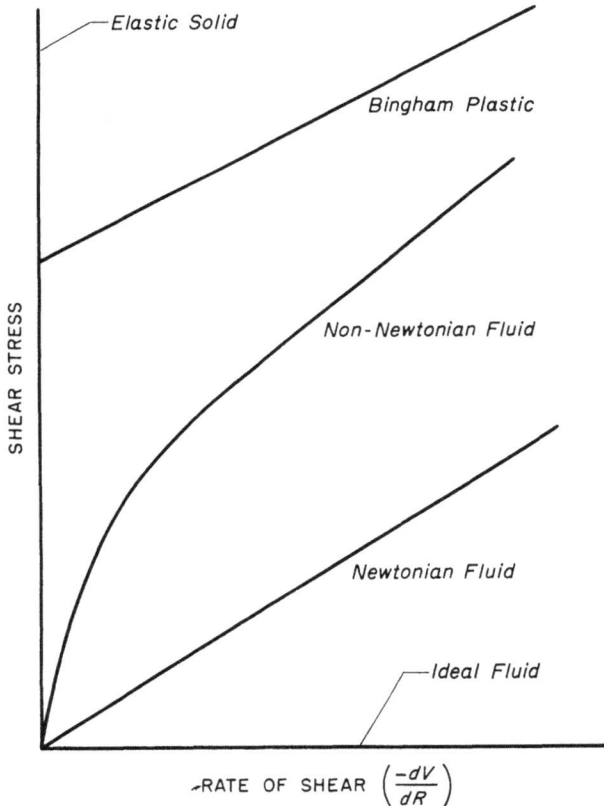

FIG. 1. RELATIONSHIP OF RATE OF SHEAR TO SHEAR STRESS
IN SOME RHEOLOGICAL CLASSES OF SUBSTANCES

explanation for these observations is somewhat uncertain but it may be that the fluids at rest consist of densely packed particles (colloidal micelles, etc.) with the intervening liquid-filled spaces being at a minimum proportion of the total volume. At low shear, the small amount of fluid acts as a lubricant, facilitating the movement of the particles over one another. When the shear increases, the voids become larger and the amount of fluid is insufficient to lubricate fully the particles. There are not many dilatant foods. Pryce-Jones (1953) described the discovery of certain honeys which possess the property of dilatancy. He ascribed this behavior to the presence of dextrans.

The time-dependent non-Newtonians include those fluids whose apparent viscosities vary not only with changes in shear stress but with time as well. According to Metzner, these substances may be examples of pseudoplastic or dilatant systems in which the molecular changes are relatively slow in occurring. Rheopectic fluids, in which shear stress increases with time of shear at a constant shearing rate, may be based on molecular arrangements similar to those causing dilatancy. Thixotropic fluids, in which the shear stress decreases with the time of shear at a constant shear rate, may be the result of the same type of system that causes pseudoplasticity. At any given time, a thixotropic fluid can be regarded as pseudoplastic and rheopectic fluids can be considered dilatant. Many sauces can be classified as thixotropic or rheopectic. Pryce-Jones (1953) has described thixotropic honeys and he attributed their behavior to the presence of a small amount of protein of unusual (for honey) type.

Although a small degree of order is possible in certain of the fluids described above, they do not possess sufficiently strong intermolecular forces to enable them to maintain a given shape in opposition to gravitational forces, i.e., they flow. Relatively strong bonds between molecules (or atoms, in the case of crystals) give solids enough mechanical strength and rigidity so that they can maintain a fixed shape in spite of the application of a considerable amount of stress, and they possess elasticity which enables them to recover a part of the deformation when stress is removed. The problems involved in measuring and describing the texture of solid foods are much more complex than those involved in the study of fluids and have received somewhat less theoretical treatment.

Elasticity and plasticity are two important elements of texture peculiar to solids. Bingham's description (1922) of these attributes is still very cogent.

If a perfectly elastic solid be subjected to a shearing stress a certain strain is developed which entirely disappears when the stress is removed. The total work done is zero, the process is reversible, and viscosity can play no part in the movement. This is not a case of flow but of elastic deformation. If a body which is imperfectly elastic as regards its form be subjected to a shearing stress, it will be found that a part, at least, of the deformation will remain long after the stress is removed. In this case work has been done in overcoming some kind of internal friction . . . . We may now define plasticity as a property of solids in virtue of which they hold their shape permanently under the action of small shearing stresses but they are readily deformed, worked, or molded, under somewhat larger stresses.

Hardness, an important component of food texture, is not considered a fundamental property of solids but as a composite property dependent upon the elastic moduli, the elastic limit, etc.

Tensile stress and tensile strain are terms used to describe the elastic properties of solids. Force per unit area of cross section of an extended body, say a muscle fiber, is defined as the tensile stress. The ratio of the change in length resulting from an increase in force is called the tensile strain. Hooke's law states that the stress is proportional to the strain for elastic bodies. However, if the stress is increased beyond a certain value, the material will not return to its original size or shape after the stress is removed. This value is called the elastic limit.

The ratio of the tensile stress to the tensile strain is called Young's modulus:

$$\Upsilon = \frac{\text{tensile stress}}{\text{tensile strain}} = \frac{F/A}{\Delta L/L} = \frac{FL}{A\Delta L}$$

Young's modulus is not a constant for many foods. Instead, it tends to exhibit elastic lag or hysteresis. The ultimate strength of a substance is defined as the maximum stress applied in rupturing it.

Particle size and particle size distribution are other important texturally-related properties of solid foods. These attributes may affect perceived texture not only in the original form of the food but as the dimensions of the fragments obtained during chewing. Particle size and distribution can be measured in many ways. Sieving tests and microscopic measuring techniques are the simplest. Sedimentation rates of suspended particles and gas permeability of compacted masses are used to give indications of average particle size. The Coulter counter has been adapted to the measurement of particle size. It records the change in dielectric properties across an orifice through which a suspension of particles is passing. From these data, the average dimensions of individual particles can be determined.

Metzner has stated that a complete measurement of food texture in a scientific and qualitative way can ultimately be based on thorough rheological tests together with particle size and distribution measurements.

More complete explanations of the properties of solids can be found in any physics textbook. The brief discussion included here is considered to be the minimum necessary to provide a background for the section on measuring devices and to define the technical terms which will be used subsequently.

Methods for Measuring Texture Of Liquids.—Determination of the rate of flow through a capillary tube, of the rate of fall of a sphere through the liquid, of the resistance offered to the rotation of an agitator immersed in the liquid, or of the torque transmitted through the liquid from a moving disc or cylinder are important ways by which viscosity can be measured. The capillary viscosimeter in one or the other of its many forms is probably the most frequently used piece of apparatus, and it utilizes the first of these principles. Fig. 2 illustrates some common types of simple capillary viscometers.

Poiseuille's law governs the flow of liquids through capillaries. Reduced to equation form, the law is: $\eta = (Pr^4/8vl)t$, where v denotes the volume of a liquid of viscosity η flowing through a tube of length l and

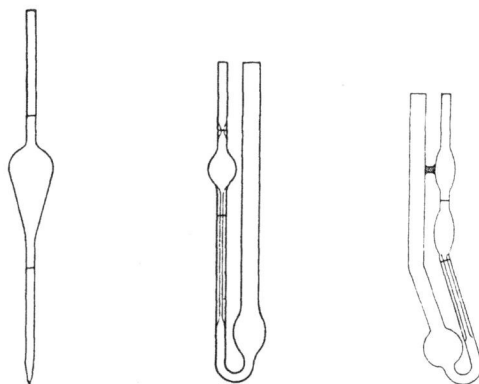

FIG. 2. THREE SIMPLE TYPES OF VISCOMETERS
Left—Dudley pipette. Center—Ostwald capillary viscometer. Right—Ostwald-Cannon-Fenske capillary viscometer.

radius r in the time t under a difference of pressure P. The law is valid only for laminar flow and the relationship does not hold if there is turbulence in the column of liquid. For any system there is a critical pressure above which turbulence will appear. The critical pressure is determined both by the geometry of the system and the characteristics of the fluid.

When two liquids are compared in the same system, under the same head of fluid, many of the terms in the above equation cancel out and the time of outflow from the capillary can be shown to be dependent upon the respective densities and coefficients of viscosity. That is, $\eta_1/\eta_2 = d_1t_1/d_2t_2$. From these considerations, the important concept of relative viscosities has been derived. If a liquid of known absolute viscosity, such as water, is selected as the standard, the preceding equation may be used to calculate very simply the relative viscosity of an unknown liquid examined under the same conditions. The absolute viscosity of the unknown may then be determined, if necessary, by multiplying the relative viscosity of the unknown by the absolute viscosity of water at the same temperature. When water is used as the standard, the preceding equation simplifies to: $\eta_1/1 = d_1t_1/d_wt_w$, where η_1, d_1, and t_1 denote the viscosity, density, and time of flow, respectively, of the experimental liquid and the corresponding sub-w terms indicate the equivalent factors for water.

An Ostwald viscometer in its simplest form is illustrated in Fig. 2, together with the Ostwald-Cannon-Fenske modification which is probably the most popular variation in use today. In making a viscosity determination with the latter instrument, a selected volume of liquid is placed in the left arm of the viscometer and suction is applied to the right arm until the meniscus of the liquid rises some distance above the upper mark on the bulb. The vacuum is then released and the time required for the meniscus to pass from the upper to the lower mark on the bulb is noted. When the instrument has been properly calibrated with distilled water, the relative viscosity of another liquid may be calculated as previously described. Determination of absolute viscosities with satisfactory precision is difficult, because of problems encountered in obtaining a uniform bore of known radius. For accurate results, it is imperative to place both the viscometer and the supplies of liquids in a constant temperature bath.

Capillary instruments are the most versatile of the viscometers. Their chief disadvantages result from the difficulties involved in cleaning and assembling them.

Since most food materials are non-Newtonian, a figure representing their true viscosity cannot be secured through use of capillary viscometers or any other instrument. It is common practice to obtain a so-called apparent viscosity for these substances by using a shear rate selected to fall within a range of practical interest. In spite of the theoretical objections to "apparent viscosity," the data secured in this manner frequently have proved to be very useful for predicting consumer preference and in control work.

According to Eolkin (1957), the rate of change of "viscosity" with changes in shear rate is an indicator of consistency or plastic flow which is valuable for many practical purposes. He designed an instrument intended to measure the viscosity of purées at two different shear rates simultaneously. Data yielded by this "Plastometer" were shown to be correlated with consistency estimations based on the Bostwick Consistometer, a popular control device (for a discussion of the latter instrument, see Davis et al. 1954). See Fig. 3.

A diagram of the Eolkin Plastometer is shown in Fig. 4. A purée is circulated through the instrument in the direction indicated by the arrows. Tubes A, B, C, and D are sized so as to be identical in flow resistance to a Newtonian fluid, which will, therefore, give a zero reading. The cross-bridge differential, $P_2 - P_4$ indicates the difference in flow resistance of the purée in two different shear zones. Under constant flow conditions, this difference is proportional to the viscosity differential and

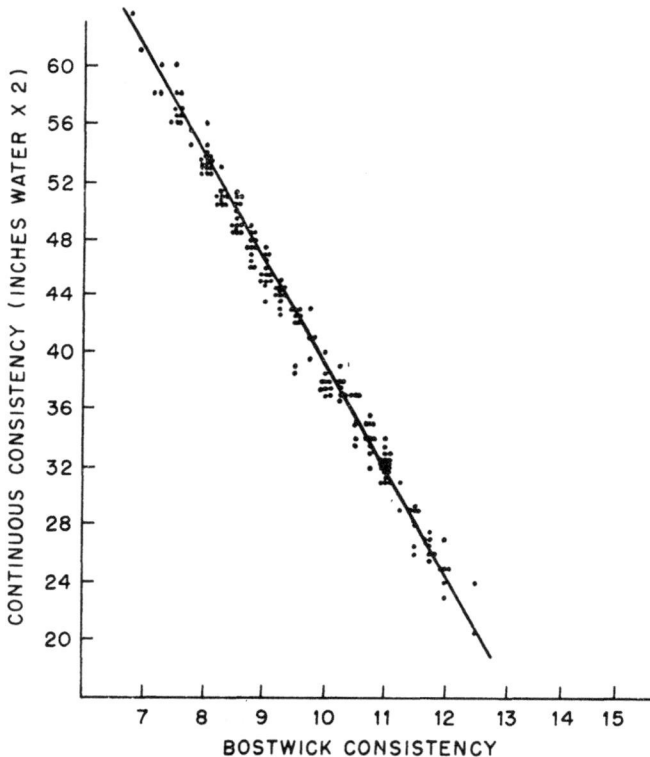

FIG. 3. CORRELATION OF BOSTWICK AND PLASTOMETER MEAS-
UREMENTS ON APRICOT PURÉE, ACCORDING TO EOLKIN (1957)

to the slope of the viscosity vs. shear rate curve. Further description of the Plastometer and theory of its application can be found in the original article. Fischer and Porter Co. have manufactured a commercial version of the Plastometer.

The Hoeppler rolling sphere viscometer has been used to determine the viscosities of milk, syrups, and other foods. In essence, this and similar devices measure the time required for a sphere (e.g., a metal ball such as a ball bearing) to fall a fixed distance in a tube filled with the liquid under investigation. By varying the weight of the sphere and its diameter relative to the diameter of the column of liquid, fluids of widely divergent viscosities can be measured. These devices are particularly useful for measuring rather viscous substances such as syrups. The duration of fall can be measured electronically if the fluid is too opaque to permit visual observation.

The theoretical justification for falling ball viscometers lies in Stoke's law relating the frictional force acting on a moving sphere to the viscosity

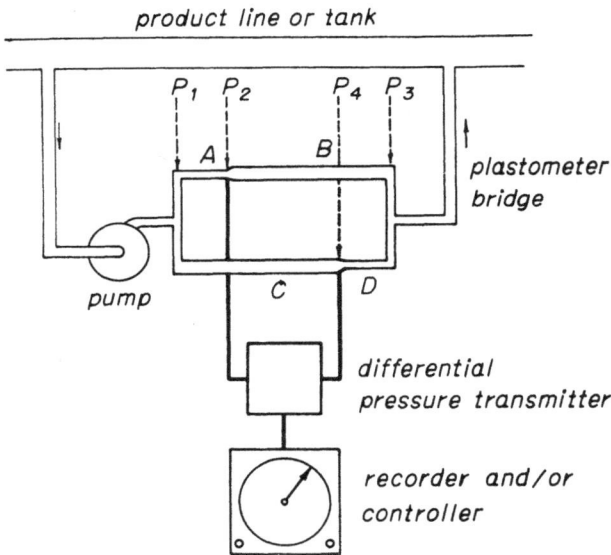

FIG. 4.   THE EOLKIN PLASTOMETER

of the liquid. Corrections must be applied for the influence on the viscosity of the wall and the bottom of the tube. For a small sphere of radius r falling axially through a viscous fluid in a cylindrical tube, the complete expression is:

$$\frac{1}{t}\left(1 + 2.4\,\frac{r}{R}\right)\left(1 + 33\,\frac{r}{h}\right) = 2gr^2\left(\frac{d_1 - d_2}{9\eta}\right)$$

when R = radius of the cylinder, h = height of the liquid, g = acceleration of gravity, d_1 = density of the sphere, and d_2 = density of the liquid.

If the same instrument (sphere and tube) is used for two different liquids, the equation can be reduced to: $\eta = K(d_1 - d_2)t$. To determine the tube constant K for a given apparatus, it is necessary to measure the time of fall through a liquid whose density and viscosity are known. If only the relative viscosity is required, the tube constant cancels out and the following expression may be used:

$$\frac{\eta}{\eta_s} = \frac{(d_1 - d_2)}{(d_1 - d_s)} \frac{t}{t_s}$$

the subscript s referring to the standard liquid.

Rotational viscometers which measure the drag on a powered agitator as an indication of the viscosity or consistency of the immersing fluid have been used in many investigations in the food field and are more popular, as control instruments, than capillary viscometers. In order to be of value in determining the "texture" of non-Newtonian fluids, these instruments must be capable of making measurements over a wide range of shear rates (to secure data relatable by some factor to texture, one must first determine the shear rate at which the texture will be subjectively evaluated and then measure comparable products at this shear rate). Many viscometers which are not capable of having their shear rates readily altered can still be useful in measuring the viscosity of substantially Newtonian fluids such as sugar syrups, skim milk, certain beverages, and the like.

Rotational viscometers of the cone-and-plate and the bob-and-cup types are useful in food investigations because they permit the study of fluids over a wide range of shear rates. In cone-and-plate viscometers, the fluid is sheared in a small gap between a cone and a flat plate. According to Metzner (1961), in units designed so that the apex of the cone just touches the surface of the plate, the shear rate is uniform throughout the entire fluid sample and is given by $u' = 2\pi N/\alpha$. Shearing stress is obtained from the torque measurement by applying the equation $T = 3t/2\pi R^3$.

In cup-and-bob viscometers, the fluid sample is sheared in the annular space between a cylindrical bob and a cup. When the ratio of the diameters of the cup to the bob is very large, the shear rate at the bob varies in accordance with the relation $u'_i = 4\pi N/n''$ for all Newtonian fluids. The shearing stress for these instruments is given by the equation $T_i = 2t/\pi D_i^2 L$, whether the annular space is broad or narrow. End effects in cup-and-bob viscometers are important sources of error, but a correction factor may be determined by calibrating the instrument with a fluid which is Newtonian within the range of interests.

The MacMichael viscometer is a widely used cup-and-bob instrument of the "narrow-gap" type, while the Brookfield Synchro-Lectric viscometer is of the wide-gap variety. One version of the latter is illustrated in Fig.

5. In this instrument, the "cup" is any container of sufficiently great diameter. The Brookfield device measures the drag produced on a spindle rotating at a constant definite speed while it (the spindle) is immersed in the material being tested. A series of rotors of various diameters and calibrated in centipoises is provided with the instrument. Three models differing in the speeds and springs are available so that a range of viscosity up to 8,000,000 centipoises can be measured. Accuracy of one per cent of full scale is claimed when the device is used

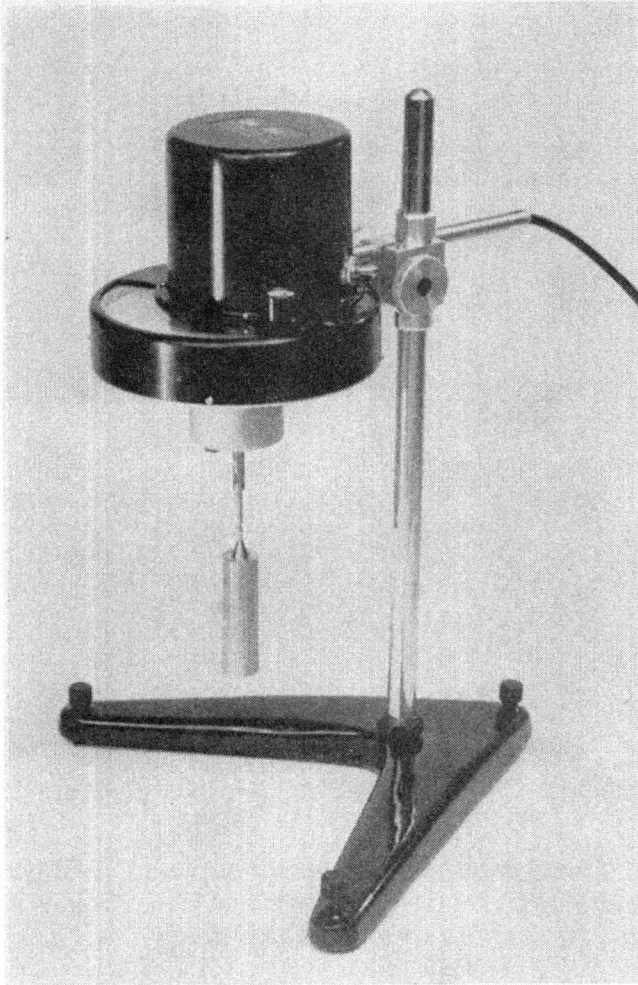

FIG. 5. THE BROOKFIELD SYNCHRO-LECTRIC VISCOMETER

in any container not less than two and three-fourths inches in diameter. Since the Brookfield device allows shear rate to be varied within a wide range, the data secured with non-Newtonian fluids can be valuable aids in assessing the textural qualities of these substances.

The MacMichael viscometer has been used in numerous food laboratories for control and investigative purposes. It records the torque

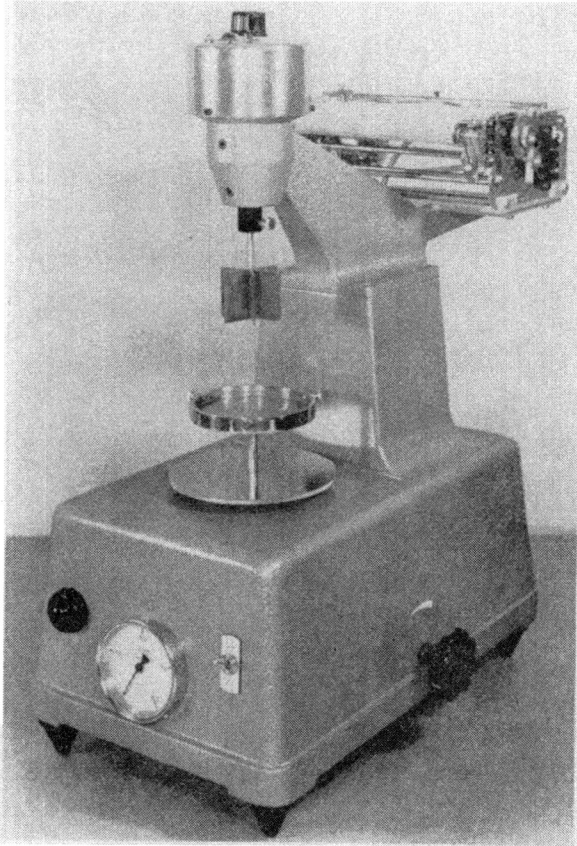

FIG. 6. THE BRABENDER VISCO-CORDER, WITH CUP REMOVED

transmitted from a rotating bowl containing the liquid to a disc suspended by a calibrated torsion wire. Provision is made for rotating the bowl at constant speed. With Newtonian fluids, the disc assumes a constant angular orientation (indicated by a pointer attached to the torsion wire) at which the forces exerted on it by the fluid and by the wire are equal.

Meaningful measurements can be made on non-Newtonian fluids by varying the bowl speed to yield measurements over a range of shear rates.

Readings on the MacMichael viscometer are taken in arbitrary degrees, but, by standardizing the device against solutions of known absolute viscosities, the results can be interpreted in terms of centipoises.

Other rotational viscometers, such as the Stormer, the Fisher electro-viscometer, the Brabender amylograph-viscograph, the Corn Industries Recording Viscometer, and the V. I. Viscograph (Hofstee and de Willigen 1953) measure the torque transmitted through liquid from a rotating

FIG. 7. THE BROOKFIELD VISCOME-TRAN

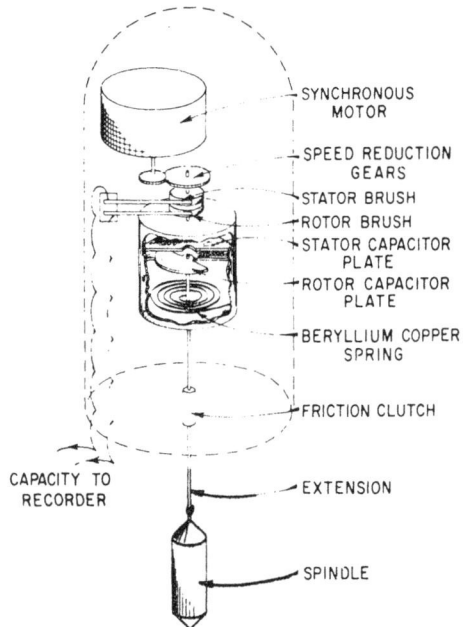

- SYNCHRONOUS MOTOR
- SPEED REDUCTION GEARS
- STATOR BRUSH
- ROTOR BRUSH
- STATOR CAPACITOR PLATE
- ROTOR CAPACITOR PLATE
- BERYLLIUM COPPER SPRING
- FRICTION CLUTCH

CAPACITY TO RECORDER

- EXTENSION
- SPINDLE

bowl which may be equipped with vertical rods or paddles to increase the energy transmitted to the fluid, or measure the energy required to turn paddles at a constant speed. In either case, the theoretical treatment is very difficult, and the data secured cannot be readily transposed into absolute units. Some of these instruments are not adjustable for different cup speeds (shear rates) and so are of doubtful utility in studying non-Newtonian behavior, even though many of them are employed for this purpose. Fig. 6 illustrates the Brabender Visco-Corder, which does have provision for changing the bowl speed. In this photograph, the cup has been removed.

Rotational instruments can be used to continuously monitor process fluids. A good example of such a viscometer is the Brookfield Viscometran, illustrated in Fig. 7. By using properly designed feedback controls it is possible to adjust consistency automatically in many food processes.

The time required for a given amount of liquid to flow through an orifice of known size is an indication of viscosity or consistency. In practice, cups of given volume with orifices of accurately determined diameter have been used, for example, in the petroleum industry. The values obtained in this manner are often meaningful for control purposes, but are not useful in research. It is doubtful if data secured through the use of such "viscometers" could have much meaning so far as texture is concerned.

The rheological characteristics of time-dependent systems, such as thixotropic and rheopectic fluids, cannot be adequately characterized by single measurements of any type. The usual procedure is to describe these systems in terms of the change in shear stress with time or by the area of the hysteresis loops between curves obtained by progressively increasing and then decreasing the shear rates. For time-dependent fluids, Metzner (1961) recommends use of rotational viscometers having small clearances between the moving surfaces to reduce variations in shear rate in different parts of the sample.

Of Solids.—The distinction between fluids and solids is clearly pointed out in this excerpt from a paper by Daugherty (1961):

A *fluid* is a substance that, no matter how viscous, will yield in time to the slightest stress; whereas a *solid*, no matter how plastic, requires a certain magnitude of stress before it will flow. With a fluid, tangential stresses between adjacent particles are proportional to the velocity of deformation and disappear when motion ceases; whereas with a solid, the tangential stresses remain and tend to restore the body to its original figure.

A solid always possesses a certain amount of elasticity as well as plasticity.

When an object is subjected to a set of forces which balance internally so that no change in size or shape results, the body is said to be "stressed." When the object is distorted by the forces, it is said to be "strained." The components of stress and the components of strain are related according to Hooke's law. For a complete discussion of Hooke's law and related subjects, the reader is referred to Timoshenko and Goodier (1951) or Alfrey (1948). These relationships are fundamental influences on texture effects in solid foods.

The behavior of viscoelastic systems is frequently diagrammed by rheologists in a manner suggesting that it is equivalent to the response of

a system of coupled dashpots and springs (Alfrey 1948). Other mechanical analogs may be introduced to explain more complicated systems
such as time-dependent bodies. A typical result is the scheme presented
by Shimizu *et al.* (1958) to explain the behavior of cooked noodles under
constant load. Their diagram is reproduced in Fig. 8. The instantaneous
elasticity of the substance is represented by the elastic spring G_1, the
retarded elasticity by G_2 coupled with a damping viscosity η_2, and the
pure viscosity component by η_3. As indicated by Alfrey (1948):

The relative importance of the elastic mechanism of response and the flow
mechanism obviously depends not only upon the magnitude of G [modulus of
rigidity or modulus of elasticity in shear] and η but also upon the timescale
of the experimental investigation.

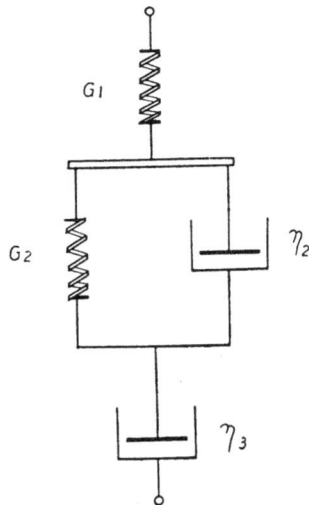

FIG. 8. MECHANICAL ANALOG OF THE RHEOLOGI-
CAL BEHAVIOR OF COOKED NOODLES UNDER CON-
STANT LOAD

Devices for measuring the components of texture of solid foodstuffs
have employed a great variety of empirically-suggested approaches. Instruments such as the Delaware jelly tester or the Bloom gel-strength
apparatus apply increments of weight to the substance and measure the
deformation. Penetrometers using plungers of many different shapes
and sizes have been used to assess some combination of elasticity and
plasticity. Cutting blades moved by a fixed force or at a fixed rate of
speed are popular for testing the texture-associated component of fibro-
sity. Extension-type testers have been applied to single muscle fibers.
Grinders, mixers, and extrusion equipment have been equipped with
force indicating means and used on many kinds of food materials. Many
of these devices are discussed in detail in the following chapters.

Chemical methods sometimes yield worthwhile correlations with subjective texture evaluations. The alcohol insoluble solids content of fruits and vegetables has been shown to be an indication of their maturity and thus of their texture. Fiber tests involving weighing of the residue left after chemical digestion or mechanical disintegration of less resistant components provide data useful in predicting consumer judgments of texture quality. Moisture content is closely related to texture in many kinds of foodstuffs, especially where crispness is the desired quality. Determination of the relative and/or absolute amounts of the various pectic fractions often provides a rough idea of the ripeness and softness of fruits. On the other hand, starch and protein contents have not been found to be important indicators of texture quality even though they form some of the structural elements of foods. Fat in meats is correlated at a rather low level with tenderness.

BIBLIOGRAPHY

ALFREY, T., JR. 1948, Mechanical Behavior of High Polymers. Interscience Publishers, New York.

BAKER, G. A., AMERINE, M. A., ROESSLER, E. B., and FILIPELLO, F. 1960. The nonspecificity of differences in taste testing for preference. Food Research 25, 810–816.

BINGHAM, E. C. 1922. Fluidity and Plasticity. McGraw-Hill Book Co., New York.

BLOOM, O. T. 1925. Penetrometer for testing the jelly strength of glues, gelatins, etc. U. S. Pat. 1,540,979. June 9.

BOGGS, M. M., and HANSON, H. L. 1950. Analysis of foods by sensory difference tests. Advances in Food Research 2, 219–258.

BOGGS, M. M., VENSTROM, D. W., HARRIS, J. G., and SHINODA, S. 1960. Performance of flavor judges in long term studies with frozen cauliflower and spinach. Food Technol. 14, 366.

BRADLEY, R. A., and KRAMER, A. 1957. Addenda to a quick, rank test for significance of differences in multiple comparisons. Food Technol. 11, 412–415.

CAUL, J. F., CAIRNCROSS, S. F., and SJÖSTRÖM, L. B. 1958. The flavor profile in review. In Flavor Research and Food Acceptance. Reinhold Publishing Co., New York.

CHARM, S. 1960. Viscometry of non-Newtonian food materials. Food Research 25, 351–362.

DAUGHERTY, R. L. 1961. Fluid properties. In Handbook of Fluid Dynamics. Edited by V. L. Streeter. McGraw-Hill Book Co., New York.

DAVIS, R. B., DeWEESE, D., and GOULD, W. A. 1954. Consistency measurements of tomato purée. Food Technol. 8, 330–334.

DETLEFSEN, G.-R. 1958. Development of a product. In Flavor Research and Food Acceptance. Reinhold Publishing Corp., New York.

DYKSTRA, O., JR. 1960. A note on the analysis of consumer preference data. Food Technol. 11, 314–315.

EOLKIN, D. 1957. The plastometer—a new development in continuous recording and controlling consistometers Food Technol. 11, 253–257.

GORTNER, R. A., GORTNER, R. A., JR., and GORTNER, W. A. 1949. Outlines of Biochemistry. John Wiley and Sons, New York.

GRIDGEMAN, N. T. 1958. Psycophysical bias in taste testing by pair comparison, with special reference to position and temperature Food Research 23, 217–220.

GRIDGEMAN, N. T. 1961. A comparison of some taste-test methods. J. Food Sci. *26*, 171–177.

HARVEY, H. G. 1953. The rheology of certain miscellaneous food products. *In* Foodstuffs: Their Plasticity, Fluidity, and Consistency. Edited by G. W. Scott Blair. Interscience Publishers, New York.

HOFSTEE, J., and DE WILLIGEN, A. H. A. 1953. Starch. *In* Foodstuffs: Their Plasticity, Fluidity, and Consistency. Edited by G. W. Scott Blair. Interscience Publishers, New York.

HOWER, R. K. 1958. Flavor testing in a baking company. *In* Flavor Research and Food Acceptance. Reinhold Publishing Corp., New York.

KRAMER, A. 1956. A quick rank test for significance of difference in multiple comparisons. Food Technol. *10*, 391–393.

KRAMER, A. 1960. A rapid method for determining significance of differences from rank sums. Food Technol. *14*, 576–581.

KRAMER, A., MURPHY, E. F., BRIANT, A. M., WANG, M., and KIRKPATRICK, M. E. 1961. Studies in taste panel methodology. J. Agr. Food Chem. *9*, 224–228.

MERRILL, E. W. 1956. A coaxial-cylinder viscometer for non-Newtonian fluids. J. Instr. Soc. Am. *3*, 124–130.

METZNER, A. B. 1956. Non-Newtonian technology: Fluid mechanics, mixing, and heat transfer. Advances in Chem. Eng. *1*.

METZNER, A. B. 1961. Flow of non-Newtonian fluids. *In* Handbook of Fluid Dynamics. Edited by V. L. Streeter. McGraw-Hill Book Co., New York.

MORSE, R. L. D. 1951. Rationale for studies of consumer food preferences. Advances in Food Research *3*, 386–427.

MURPHY, E. F., COVELL, M. R., and DINSMORE, J. S. 1957. An examination of three methods for testing palatability as illustrated by strawberry flavor differences. Food Research *22*, 423–430.

PERRY, J. H. 1950. Chemical Engineers Handbook. McGraw-Hill Book Co., New York.

PERYAM, D. R. 1958. Sensory difference tests. *In* Flavor Research and Food Acceptance. Reinhold Publishing Corp., New York.

PILGRIM, F. J. 1961. Private communication.

PRYCE-JONES, J. 1953. The rheology of honey. *In* Foodstuffs: Their Plasticity, Fluidity, and Consistency. Edited by G. W. Scott Blair. Interscience Publishers, New York.

RIETZ, C. A. 1961. A Guide to the Selection, Combination, and Cooking of Foods. Avi Publishing Co., Westport, Conn.

SATHER, L. A., and CALVIN, L. D. 1960. The effect of number of judgments in a test on flavor evaluations for preference. Food Technol. *14*, 613–615.

SCHACHAT, R. E. and NACCI, A. 1960. Transistorized bloom gelometer. Food Technol. *14*, 177–118.

SCOTT BLAIR, G. W. 1958. Rheology in food research. Advances in Food Research *8*, 1–61.

SHIMIZU, T., FUKAWA, H., and Ichiba, A. 1958. Physical properties of noodles. Cereal Chem. *35*, 34–46.

TIMOSHENKO, S., and GOODIER, J. N. 1951. Theory of Elasticity. McGraw-Hill Book Co., New York.

Types of Texture in Foods

CHAPTER 3

Liquid Foods

INTRODUCTION

This chapter contains discussions of only those foods which flow readily at room temperature. Gels and fatty foods which liquefy near body temperature are considered in subsequent chapters.

Soups, syrups, juices, some sauces, and milk and other beverages are among the important types of foods which can be fitted into this category. In addition, there are many compound foods which are predominantly liquid but which include perceptible particles, such as those beverages which contain small pieces of fruit (citrus "squashes") and most purées and soups.

Liquid foods which are essentially Newtonian fluids may be considered to have their texture or "mouthfeel" adequately described by their viscosity. However, most liquid foods are probably non-Newtonian. Nonetheless the measurement of viscosity or apparent viscosity has become an important method for estimating the quality of both classes of these foods. The concept of viscosity and its measurement were discussed in the preceding chapter.

PRODUCTS

Milk

The texture or viscosity of milk is an important factor influencing acceptance even though the effect may be obscured because of the relatively small variation in viscosity of different commercial samples. As stated by Jenness and Patton (1959):

In many instances "richness" in the eyes of the consumer is indicated by the viscosity of a product. On the other hand, consumer taste in the matter of viscosity may show great variation. For example, some sections of the United States prefer a highly viscous chocolate milk, whereas others desire a relatively "thin" product. . . . The viscosity of table cream and buttermilk is an obvious consideration in their acceptance by the consuming public.

Homogeneous whole milk seems to behave essentially as a Newtonian fluid in spite of the presence in it of two distinct phases. However, viscosity measurements on whole milk are complicated by the separation of cream, which does not behave in accordance with Newtonian principles. Cox (1952) and Scott Blair (1953) have published reviews of recent investigations of the viscosity of milk.

Cox *et al.* (1959) reviewed previous work and showed clearly that serious differences existed between the results obtained by different investigators in the field. However, the analyses surveyed by them did give quantitative confirmation of the general knowledge that, within the ranges of composition usually found in normal milk, the effective viscosity increased as the fat percentage increased in milks with constant solids-not-fat. Similarly, viscosity increased with solid-not-fat percentage for milks which had the same percentage of fat.

Milk proteins play the most important role in the viscosity of skim milk, the approximately five per cent of lactose having little effect. Of the classes of protein present in milk, casein has by far the most influence on viscosity. At the usual temperatures of consumption, skim milk is less viscous than whole milk.

Cox (1952) has stated that the bulk of published evidence indicates that there is a virtually linear decrease in the viscosity of milk with increasing temperature, at least up to 86°F., with a subsequent lessening of the rate of decrease. The data on which he based these conclusions are presented in Table 1.

Whitnah *et al.* (1956) found that homogenization of whole pasteurized milk at pressures above 300 lbs. per sq. in. caused highly significant changes in viscosity. Linear and quadratic relationships between pressure and viscosity change were observed. Pressures of 1, 1.5, 2, 3, and 3.5 thousand lbs. per sq. in. caused average increases in viscosity of 7.1, 9.2, 11.9, 13.7, and 15.0 per cent, respectively. The relative increases in viscosity were not materially affected by temperature changes over the range 39° to 120°F. At pressures lower than about 300 lbs. per sq. in., linear trends were usually significant if the temperature was above 73°F. At lower temperatures there were no significant linear or quadratic trends. Cubic trends sometimes indicated an increase in viscosity at pressures of about 300 lbs. per sq. in. after an initial decrease which began at pressures of about 15 lbs. per sq. in.

Bateman and Sharp (1928) had previously noted that homogenization caused an increase in the viscosity of milk and had also indicated that pasteurization led to a slight fall in the viscosity. They state that the viscosity of skim milk increases with age. It is known that the viscosity of pasteurized milk (homogenized or non-homogenized) increases with age as a linear function of the storage time. This may be due to fat clustering, equilibration of the milk salts, or the action of bacteria.

Increases in viscosity upon homogenization are probably the result of increases in the area of the fat surface as the globules are fragmented, and of the amount of protein bound by the fat. The extent to which the fat globules are clumped together doubtless has some effect.

TABLE 1

ABSOLUTE VISCOSITIES OF SIX HOMOGENIZED MILK SAMPLES AT DIFFERENT TEMPERATURES[1]

Sample No. 1		Sample No.2		Sample No. 3		Sample No. 4		Sample No. 5		Sample No. 6	
Temp.[2]	Visc.[3]	Temp.	Visc.	Temp.	Visc.	Temp.	Visc.	Temp.	Visc.	Temp.	Visc.
67	1.98	68	1.99	68	2.06	72	1.83	68	2.01	69	1.92
86	1.47	86	1.53	87	1.50	87	1.48	82	1.52	91	1.36
105	1.11	108	1.13	104	1.19	103	1.16	105	1.13	103	1.18
122	0.92	120	0.98	121	0.98	121	0.97	119	0.96	124	0.92
140	0.76	138	0.82	140	0.85	138	0.86	139	0.79	154	0.70
158	0.66	156	0.70	159	0.71	157	0.77	157	0.68	...	...
173	0.60	175	0.63	175	0.64	177	0.61	174	0.61	...	...

[1] Adapted from data of Caffyn (1951).
[2] Temperatures are in degrees Fahrenheit.
[3] Viscosities are in centipoises.

Table Syrups

The most important table syrups consumed in the United States are corn syrup, honey, molasses, and, especially, the flavored and colored blends of sucrose and maple syrup. Viscosity is recognized by consumers and producers as an important indicator of quality in table syrups.

Honey is a concentrated solution of fructose and glucose with traces of other carbohydrates, proteins, minerals, vitamins, etc. Pryce-Jones (1953) has published a rather extensive review of the rheology of honeys. Most of these products appear to act as simple Newtonian fluids. Their viscosity at any given temperature is largely determined by the sugar concentration.

There are a few remarkable exceptions. Heather (*Calluna vulgaris*) honey, one of the important European honeys, is thixotropic, or, more correctly, is a "false-body" system as defined by Pryce-Jones. The gelatinization which occurs when heather honey is allowed to rest undisturbed is due to the presence of about 1 to 2 per cent of a characteristic protein. Heating to about 149°F. accentuates the gelling behavior of the honey by denaturing the protein.

Manuka (*Leptospermum scoparium*) honey from New Zealand is also markedly thixotropic. Gelling in this honey is also a result of the presence of about one per cent of protein in the substance.

Pryce-Jones states that he has observed a large number of honeys which are dilatant, that is their viscosity increases with increasing shear beyond a critical minimum value. Some samples of honey from *Eucalyptus ficifolia* were especially notable in this respect. All of the honeys exhibiting dilatancy contained dextran, $(C_6H_{10}O_5)_n$, in which n is in the region of 8,000. The dextran is present in amounts as high as 7.2 per cent of the total weight of the honey, and it is probably synthesized by the micro-organism *Leuconostoc mesenteroides*. After the dextran was removed by treatment of the honey with acetone, the honey behaved like a true Newtonian fluid. Furthermore, the peculiar rheological effects could be reproduced in, e.g., clover honey, by the addition of dextran.

Sucrose syrups and syrups produced by blending cane or beet with maple syrups behave as simple Newtonian fluids. Their texture is a function of the sugar concentration. A linear relation exists between the logarithm of the viscosity and the concentration of soluble solids when the latter is expressed as the ratio of the weight of sugar dissolved per unit weight of water. Lipscomb (1956) reported an extensive study of the viscosities of single sugars and mixtures of sugars at various concentrations. Table 2 reports the viscosity of three different concentrations of sucrose at nine different temperatures representing the range throughout which consumption might occur.

Maple syrup must not contain more than 35 per cent water in order to meet legal standards. The solids are predominantly sucrose with a few per cent reducing sugars. It is not likely that the non-carbohydrate components make a significant contribution to the viscosity of the product. Maple syrup is, therefore, a Newtonian fluid for all practical purposes and an adequate measure of its texture or viscosity can be obtained by measuring the rate of flow at any shear rate. Certain micro-organisms can alter the characteristic texture to give a "stringy" quality to maple syrup (Edson 1910). This change may be similar in cause to the microbial spoilage of honey which is accompanied by the elaboration of dextrans.

TABLE 2

VISCOSITY OF AQUEOUS SUCROSE SOLUTIONS[1]
In Centipoises

Temperature, °F.	Concentration of Sucrose, Per Cent by Weight		
	20	40	60
32	3.804	14.77	238
41	3.154	11.56	156
50	2.652	9.794	109.8
59	2.267	7.468	74.6
68	1.960	6.200	56.5
77	1.704	5.187	43.86
86	1.504	4.382	33.78
95	1.331	3.762	26.52
104	1.193	3.249	21.28

[1] Adapted from data of Bingham and Jackson (1919).

Sorghum syrup is prepared, mostly in the southeastern United States, by methods reminiscent of molasses manufacture. Since most of this product is made in small plants with little or no attempts at quality control, it can vary widely in composition as compared to other table syrups. It frequently contains about 75 per cent carbohydrates, at least 95 per cent of this fraction being sucrose and reducing sugars. Consequently, it, too, is Newtonian in response. The texture is largely a function of the moisture content, small changes in the amount of water removed during manufacture having large effects on the rheological properties.

Products made by boiling down dilute sucrose solutions, maple, sorghum, and molasses syrups for example, contain amounts of invert sugar which vary according to the conditions existing during evaporation. Inversion of sucrose causes a reduction in the viscosity of a syrup, as shown in Table 3. Consequently, mild conditions, such as a pH close to neutrality, are desirable during the boiling process in order to obtain syrups of the highest viscosity.

Corn syrups, which are made by acid- or enzyme-hydrolysis of starch, consist of glucose and a very diverse collection of polymers of glucose ranging from maltose to compounds with molecular weights of several thousand. Although it is the usual commercial practice to describe a syrup in terms of its moisture content (usually indirectly, as the density in degrees Baumé) and dextrose equivalent, these data are not sufficient to establish the viscosity of the product. The large molecular weight compounds, or dextrins, exercise a disproportionately large effect on the viscosity. Special processing techniques can be used to yield syrups which have about the same dextrose equivalent but which differ considerably in viscosity. Table 4 illustrates this fact. The syrup described in the last column has been hydrolysed by a method which yields far less glucose and more maltose and saccharides with three or more glucose residues. Evidently the dextrins which are present in the latter product have a higher average molecular weight accounting for a higher viscosity observed in the latter product.

TABLE 3

EFFECT OF INVERSION ON VISCOSITY OF SUCROSE SOLUTIONS[1]

Dry Analysis, Per Cent Sugar		Viscosity at 72 °F. and 65 Per Cent Solids[2]
Invert	Sucrose	
0	100	133
10	90	127
50	50	97

[1] From Davis and Prince (1955).
[2] Viscosities expressed in centipoises.

TABLE 4

RELATIONSHIP OF VISCOSITY TO OTHER CHARACTERISTICS OF CORN SYRUPS

Extent of Conversion →	Low	Regular	Interme-diate	High	Extra High	High
Type of Hydrolysis →	Acid	Acid	Acid	Acid-Enzyme	Acid-Enzyme	Special Acid-Enzyme
Density, degrees Baumé	42	43	43	43	43	43
Total solids, per cent	78	80	81	82	82	80
Dextrose equivalent	32	42	54	64	71	43
Monosaccharides, per cent	12	19	30	39	41	6
Dissaccharides, per cent	10	14	18	33	40	43
Higher saccharides, per cent	78	67	52	28	18	51
Viscosity, poises at 100 °F.	100	125	70	50	46	145

According to the literature (Harvey 1953), corn syrups behave as simple Newtonian liquids. However, some discrepancies might be expected, especially with syrups of relatively low dextrose equivalent. Insofar as they conform to this simple viscosity pattern, their texture can be described by a single figure—the viscosity at the temperature of consumption.

Some table syrups consist of a blend of corn and sucrose syrups. According to Davis and Prince (1955), an approximation of the viscosity of such a blend can be calculated by proportioning the logarithms of the viscosities of the ingredients using the antilog as the approximate viscosity of the blend.

Fruit and Vegetable Juices and Purées

Most juices and purées obtained from fruits and vegetables are non-Newtonian as a result of the presence of suspended particles and dissolved long chain molecules. Harper (1960) indicates that most purées fall into the class of pseudoplastic fluids. As stated by Tressler and Joslyn (1954), consistency is an important quality factor in these products, but no objective methods for measuring consistency are recognized in the present Federal standards. In the U. S. Standards for Grades of Tomato Juice, consistency is said to "pertain to viscosity or separation of solids." It is certain that most consumers regard products of greater viscosity as being of better quality than juices or purées which are "thinner."

Kertesz and Loconti (1944) also indicated that consistency as a quality factor in vegetable juices has not been clearly defined and is affected by several chemical and physical factors whose relationships to the sensory characteristics of the product have not been fully elucidated. They recognized that "gross [apparent] viscosity" cannot be used as a single measurement in unclarified juice since this property is affected by the character, size, and proportion of the suspended particles. Gross [apparent] viscosity is, however, the most common method used in the evaluation of the consistency of canned tomato juice, according to Kertesz and Loconti. The rate of flow through pipettes containing an orifice of given size is frequently used for this measurement. In spite of the many theoretical objections to the determination of texture-related properties in this manner, the data appear to be valuable for control purposes. These authors recommend that the size of the pipette orifice be large in relation to the size of the particles in the juice, and so a pipette with a large orifice should be used if reproducible data are to be obtained.

Of several instruments tested by these investigators for determining apparent viscosity in tomato juice, the Stormer viscometer was found to be the most satisfactory. They also stated that a viscous serum is essential for a desirable consistency and that measurement of the relative viscosity of filtered tomato juice using an Ostwald viscometer gave a satisfactory measurement. Actually, the use of pipettes, Stormer viscometers, or other devices with fixed shear rates cannot be expected to give fully satisfactory texture ("viscosity") measurements on non-Newtonian fluids such as filtered or unfiltered tomato juice.

Whittenberger and Nutting (1957) concluded that the consistency of tomato juice depends partly on its chemical composition and partly on its physical structure. They said

A tomato is composed principally of relatively large, near-spherical flask cells which are filled with dilute sugar water or cell sap and small quantities of living matter and insoluble granules. The largest cells can barely be seen with the naked eye.

The outer wall of the cell consists of interwoven cellulose fibers impregnated with pectic compounds. The cell walls, although separated, remain capable of exerting an important, or even predominating, effect on juice consistency despite the fact that they amount to only a per cent or less by weight of the juice. Maturity and the variety of the fruit, preheating treatment, and manner of extracting can influence the quantity of cell walls in a juice.

In a later publication, Whittenberger and Nutting (1958) indicated that consistency in tomato juice was dependent primarily on the structure and composition of the cell walls. They de-emphasized the role of soluble pectins but reaffirmed and strengthened the role of insoluble solids. The cell walls were found to contain amorphous cellulose as opposed to the crystalline cellulose predominating in wood and cotton. Tomato cellulose alone formed suspensions that were thicker, at equivalent concentrations, than most of the common thickening agents.

According to Whittenberger and Nutting, the viscosity of tomato juice is kept at a relatively low level by natural and added electrolytes. Removal of such electrolytes as soluble pectins, organic acids, and mineral salts may cause the remaining juice to thicken to a semi-gel. Cell walls comprise less than six per cent of the total solids and less than 50 per cent of the insoluble solids. As electrolytes decrease, the walls swell. The viscosity will remain high in the presence of such non-electrolytes as sucrose, glycerol, and ethanol, but will decrease upon the addition of electrolytes. Homogenization increases viscosity, apparently by increasing the surface area.

Smit and Nortje (1958) published a number of observations on the consistency of tomato paste which appear to agree well with the conclusions of Whittenberger and Nutting. They manufactured paste from San Marzano and Pearson tomatoes using two different screen sizes in the finisher. Some samples were immediately heated to 180°F. after breaking while others were delayed for three hours at 70° to 80°F. between the breaking and preheating steps. They found that a larger screen size in the finisher gave a paste with a higher consistency. A delay between breaking and preheating resulted in a higher consistency, at the same level of total soluble solids, even though the calcium pectate, methoxyl content, and the comparative serum viscosity were lower than in samples which were immediately preheated before breaking. Smit and Nortje suggested that the difference in consistency was due to a qualitative difference in the insoluble substances.

Pectic substances are doubtless the predominant viscosity determining factors in apple, citrus, and strawberry (Simpson 1959) juices. Enzymes present in the fruit or resulting from microbiological activity may seriously affect the viscosity of the juices (Pollard 1958) by demethoxylating or depolymerizing the pectic substances.

Manufactured Beverages

The chief kinds of "manufactured" beverages, as opposed to fruit and vegetable juices, are carbonated soft drinks, beer, wine, and distilled alcoholic beverages. Artificially flavored and colored non-carbonated soft drinks make up another commercially important class in this category.

Rietz (1961) states the well-recognized fact that some wines have more "body" than others due to differences in amounts and kinds of suspended and dissolved substances. He also regards the effervescence of carbon dioxide from wines and carbonated beverages as contributing to the texture of these drinks.

As shown in Table 5, ethanol when present in the concentrations usually found in alcoholic beverages, increases the viscosity of aqueous solutions. The viscosity of distilled beverages should be essentially the result of the water and ethanol. Polysaccharides carried over from the malt and cereal adjuncts influence the viscosity of beer. Wine may contain pectic substances derived from the fruit in amounts sufficient to affect the body of the beverage. The addition of sucrose, corn syrup, or honey to ethanolic solutions in the preparation of liqueurs and cordials can alter the fluidity

TABLE 5

VISCOSITY OF ALCOHOL-WATER MIXTURES[1]

In Centipoises

Temperature, °F.	Concentration of Ethanol, Per Cent By Weight									
	10	20	30	40	50	60	70	80	90	100
32	3.311	5.32	6.94	7.14	6.58	5.75	4.76	3.69	2.73	1.77
41	2.58	4.06	5.29	5.59	5.26	4.63	3.91	3.12	2.31	1.62
50	2.18	3.16	4.05	4.39	4.18	3.77	3.27	2.71	2.10	1.47
59	1.79	2.62	3.26	3.53	3.44	3.14	2.77	2.31	1.80	1.33
68	1.54	2.18	2.71	2.91	2.87	2.67	2.37	2.01	1.61	1.20
77	1.32	1.82	2.18	2.35	2.40	2.24	2.04	1.75	1.42	1.10
86	1.16	1.55	1.87	2.02	2.02	1.93	1.77	1.53	1.28	1.00
95	1.01	1.33	1.58	1.72	1.72	1.66	1.53	1.36	1.15	0.91
104	0.91	1.16	1.37	1.48	1.50	1.45	1.34	1.20	1.04	0.83

[1] Adapted from data of Bingham and Jackson (1919).

to a marked degree. The resins added to some Greek wines and liqueurs doubtless increase the viscosity although data on this point are lacking.

Beverages composed of ethanol, water sucrose, and minor amounts of coloring and flavoring agents can be expected to correspond to a Newtonian model. More complicated behavior ensues when pectic substances, dextrins, or carbon dioxide are present and Newtonian viscosity cannot be expected to give a completely satisfactory explanation of the rheological properties of these beverages.

According to Gortatowsky (1955), the function of the carbohydrate in bottled drinks is that of a sweetener and provider of body or mouthfeel. In 1952, 73 per cent of soft drink bottlers used sucrose as the sole carbohydrate while 27 per cent used dextrose in a 5 to 45 per cent mix with sucrose. Most of these beverages have a pH range of 2.5 to 3.5 which is sufficiently low to cause inversion of part of the sucrose. As shown previously in this chapter, this change will cause a decrease in the viscosity of the drink.

It would appear that a simple measurement of apparent viscosity at the intended temperature of consumption would give an adequate measure of the texture of most beverages even though many of them are non-Newtonian.

Soups

Commercial canned cream style soups are thickened with starches or low-protein wheat flours. The characteristics of starch suspensions will be discussed fully in the chapter on gel-type foods. Salt and the relatively low molecular weight polymers of amino acids which are other important constituents of most soups can be expected to have only minor effects on consistency because of their low concentration. The fatty components are frequently present in large globules or masses, i.e., not as emulsions, and affect mouthfeel in a complex manner. In any case, simple viscosity measurements are not likely to give a complete picture of the sensory response to these foods.

BIBLIOGRAPHY

ASSELBERGS, E. A., FERGUSON, W. E., and MACQUEEN, E. F. 1958. Effects of sodium sorbate and ascorbic acid on attempted gamma irradiation pasteurization of apple juice. Food Technol. 12, 156–158.
BATEMAN, G. M., and SHARP, P. F. 1928. A study of the apparent viscosity of milk as influenced by some physical factors. J. Agr. Research 36, 647–674.
BINGHAM, E. C., and JACKSON, R. F. 1919. Standard substances for the calibration of viscometers. J. Wash. Acad. Sci. 7, 53–55.
BISSETT, O. W. 1958. Processing freeze-damaged oranges. Proc. Fla. State Hort. Soc. 71, 254–259.

CAFFYN, J. E. 1951. The viscosity temperature coefficient of homogenized milk. J. Dairy Research 18, 95–104.

CHARM, S. 1960. Viscometry of non-Newtonian food materials. Food Research 25, 351–362.

COX, C. P. 1952. Changes with temperature in the viscosity of whole milk. J. Dairy Research 19, 72–78.

COX, C. P., HOSKING, Z. D., and POSENER, L. N. 1959. Relations between composition and viscosity of cow's milk. J. Dairy Research 26, 182–189.

DAVIS, J. G. 1937. The rheology of cheese, butter, and other milk products. J. Dairy Research 8, 245–264.

DAVIS, J. G. 1956. Yoghurt and other cultured milks. J. Soc. Dairy Tech. 9, 69–74, 117–122, 160–165.

DAVIS, P. R., and PRINCE R. N. 1955. Liquid sugar in the food industry. Advances in Chem. Ser. No. 12, 35–42.

EDSON, H. A. 1910. The influence of micro-organisms upon the quality of maple syrup. J. Ind. Eng. Chem. 2, 325–327.

EZELL, G. H., and OLSEN, R. W. 1958. Effect of stabilization temperature on the viscosity and stability of concentrated orange juices. Proc. Fla. State Hort. Soc. 71, 186–189.

GORTATOWSKY, C. 1955. Sugar and other carbohydrates in carbonated beverages. Advances in Chem. Ser. No. 12, 70–74.

HARPER, J. C. 1960. Viscometric behavior in relation to evaporation of fruit purées. Food Technol. 14, 557–561.

HARVEY, J. G. 1953. The rheology of certain miscellaneous food products. In Foodstuffs: Their Plasticity, Fluidity, and Consistency. Edited by G. W. Scott Blair. Interscience Publishers, New York.

HOFSTEE, J., and DE WILLIGEN, A. H. A. 1953. Starch. In Foodstuffs: Their Plasticity, Fluidity, and Consistency. Edited by G. W. Scott Blair. Interscience Publishers, New York.

JENNESS, R., and PATTON, S. 1959. Principles of Dairy Chemistry. John Wiley and Sons, New York.

KERTESZ, Z. I., and LOCONTI, J. D. 1944. Factors determining the consistency of commercial canned tomato juice. N. Y. Agr. Expt. Sta. Tech. Bull. 272.

LIPSCOMB, A. G. 1956. Rheology of candy syrup and boiled candies. Fette Seifen Anstrichmittel 58, 875–879.

MERRILL, E. W. 1956. A coaxial-cylinder viscometer for non-Newtonian fluids. J. Inst. Soc. Am. 3, 124–130.

METZNER, A. B. 1956. Non-Newtonian technology: Fluid mechanics, mixing, and heat transfer. Advances in Chem. Eng. 1.

MOYER, J. C., ROBINSON, W. B., RANSFORD, J. R., LaBELLE, R. L., and HAND, D. B. 1959. Processing conditions affecting the yield of tomato juice. Food Technol. 13, 270–275.

PETERS, G. L. 1959. Effect of varying chemical compositions brought about by processing methods on serum viscosity and water retention of tomato juice. Dissert. Abstr. 19, 3098–3099.

POLLARD, A. 1958. Fermentation of diluted concentrates. Int. Fed. Fruit Juice Producers' Symposium, Bristol, 1958, 351–360.

POWRIE, W. D., and ASSELBERGS, E. A. 1956. Influence of post-harvest ripening of McIntosh apples on the yield, composition, and flavor of the juice. Canadian J. Agr. Sci. 36, 349–355.

PRYCE-JONES, J. 1953. The rheology of honey. In Foodstuffs: Their Plasticity, Fluidity, and Consistency. Edited by G. W. Scott Blair. Interscience Publishers, New York.

RIETZ, C. A. 1961. A Guide to the Selection, Combination, and Cooking of Foods. Avi Publishing Co., Westport, Conn.

ROUSE, A. H., and ATKINS, C. D. 1955. Pectin esterase and pectin in commercial citrus juices, as determined by methods used at the Citrus Experiment Station. Univ. Florida Agr. Expt. Sta. Bull. 570.

ROUSE, A. H., ATKINS, C. D., and MOORE, E. L. 1958. Chemical characteristics of citrus juices from freeze-damaged fruit. Proc. Florida State Hort. Soc. 71, 216–219.

SCHWARZ, T. W., and LEVY, G. 1957. Viscosity changes of sodium alignate solutions after freezing and thawing. J. Am. Pharm. Assoc., Sci. Ed., 46, 562–563.

SCOTT BLAIR, G. W. 1953. Rheology of milk, ice-cream mixes, and similar products. In Foodstuffs: Their Plasticity, Fluidity, and Consistency. Edited by G. W. Scott Blair. Interscience Publishers, New York.

SIMPSON, M. 1959. Viscosity as related to pectin content in strawberry juices. Ontario Hort. Expt. Sta. Products Lab. (Vineland) 1957–1958, 126–129.

SMIT, C. J. B., and NORTJE, B. K. 1958. Observations on the consistency of tomato paste. Food Technol. 12, 356–358.

TRESSLER, D. K., and JOSLYN, M. A., 1954. The Chemistry and Technology of Fruit and Vegetable Juice Production. Avi Publishing Co., Westport, Conn.

TRUSCOTT, J. H. L., and WICKSON, M. 1956. Fruit maturity as indicated by juice viscosity. Ontario Hort. Expt. Sta. Products Lab. (Vineland) Rept., 1953–1954, 105–107.

WHITNAH, C. H. 1959. The surface tension of milk. A review. J. Dairy Sci. 42, 1437–1449.

WHITNAH, C. H., RUTZ, W. D., and FREYER, H. C. 1956. Some physical properties of milk. III. Effects of homogenization pressures on the viscosity of whole milk. J. Dairy Sci. 39, 1500–1505.

WHITTENBERGER, R. T., and NUTTING, G. C. 1957. Effect of tomato cell structures on consistency of tomato juice. Food Technol. 11, 19–22.

WHITTENBERGER, R. T., and NUTTING, G. C. 1958. High viscosity of cell wall suspensions prepared from tomato juice. Food Technol. 12, 420–424.

CHAPTER 4

Gels

INTRODUCTION

A vast number of manufactured foods and many natural substances used as foods can best be described from a textural standpoint as gels and most natural foods contain gel components even though they may not be the predominant influence on texture.

Foods composed principally of certain polymers of carbohydrates, such as starch or agar, or of proteins, such as gelatin, readily form gels when hydrated to the proper level. Some of the individual foods found in this class are starch puddings, gelatin desserts, fruit jellies, tofu, cottage cheese, alimentary pastes, cooked egg white, and gum drops. There are many cooked vegetable foods which depend for their textural characteristics on the functional properties of starch gels even though they are made up of agglomerations of cells whose turgidity, cell wall structure, and intercellular adhesion affect texture to some extent. Rice, potatoes, and cereal gruels are common examples. These foods will be discussed in detail in this chapter and brief additional consideration will be given in one of the subsequent chapters to those structural elements other than gels which contribute to their texture. Foods which depend primarily upon cellular characteristics (osmotic pressure, adhesion, etc.) for their textural qualities will be discussed in Chapter 6.

PHYSICAL AND CHEMICAL DETERMINANTS OF GEL CHARACTERISTICS

Gels are commonly regarded as two-phase systems with a high degree of interface between a continuous, or at least intermeshed, system of solid material holding an aqueous (or other solvent) phase which may also be continuous or finely dispersed. The solid material is usually thought of as fibrillar in form and may consist either of strongly solvated molecules, such as proteins, or of threadlike crystals, as in gels of silicic acid or crystalline cellulose. A certain degree of structural stability is contributed by the extended skeleton of the continuous phase so that the system as a whole possesses many of the properties of a solid, e.g., rigidity and elastic response to applied distorting forces.

The fibrillar components of gels are linked together by forces which may be covalent in nature but which are usually weaker, such as hydrogen bonds. These bonds can be distorted or broken by forces applied to the gel mass. When broken, they can be re-established at other locations on neighboring molecules. These bonds are also very susceptible to changes in temperature, becoming much more labile as the temperature increases.

The amount of water (or other solvent) present in a gel is indefinite. If the gel is placed in an excess of solvent it will usually swell until a sol is formed. On the other hand, solvent can be removed by thermal methods or by applying pressures. The forces required to squeeze the water out of gels are usually quite large. Conversely, dehydrated gels can take up water against very high pressures. However, solvent is often spontaneously expressed from gels as they age. This process, called syneresis, is evidently due to the formation of additional intermolecular bonds with a consequent reduction in the number of loci available for solvent binding and to a decrease in the dimensions of the intermolecular spaces in which the solvent is contained.

Addition to the sol of molecules which compete with the solvent for bond sites profoundly affects the quality of the gel, usually in the direction of increased rigidity. Similar effects occur when ions capable of bridging electrophilic or electronegative sites on different molecules are included in the system. Calcium is often particularly effective in this regard if the gel contains a considerable number of carboxyl groups.

Most sol-gel systems can be reversed at will by changing the temperature. The temperature at which a reversible gel melts is usually higher than that at which it sets; the lag is called hysteresis. However, the melting and setting temperatures can be made virtually the same if the rates of temperature change are kept sufficiently slow.

Many gels can be made fluid by mechanical agitation (e.g., stirring) and then will regain their original semisolid condition after a rest period. This behavior is thought to be due to changes in the electric energy potentials of attraction and repulsion between the colloidal particles. A condition of equilibrium between these forces exists only when a certain distance separates the particles. When the particles are agitated, the equilibrium is disturbed and liquefaction occurs. Gels exhibiting this behavior are said to be thixotropic.

Freshly prepared gelatin gels frequently exhibit rheopexy. Rheopexy seems to be a variation of thixotropy and is observed also in bentonite suspensions and in some other inorganic systems. The term is applied to gels which set more rapidly when slowly stirred in a uniform manner but liquefy when vigorously agitated. In these cases, the increase in rate of gelation is said to result from a more rapid alignment in parallel arrays of internal phase particles having a laminar or fibrillar shape.

The gels with which the food technologist is concerned owe their properties to compounds of high molecular weight, usually branched in form, and capable of a high degree of hydration. Starches, proteins, and the polyuronic acid gels are those most commonly found to affect the texture of foods.

TYPES OF GELS FOUND IN FOODS

Because of the limited space available, it will be impossible to discuss in detail all of the many molecular species making up the gels found in foods. The approach taken in this section is to review a few of the most important gel-forming substances and then to relate the findings in these specific cases to similar types of structures.

Starch Gels

Many compounded and natural foods owe their characteristic texture to the presence of starch gels. Macaroni products, potatoes, rice, and starch puddings are examples of such foods. In some of these foods, other classes of structures contribute significantly to the perceived texture. This is true in alimentary pastes, where the gluten proteins constitute a continuous phase enveloping the starch, and in potatoes where the starch is partly contained inside cell walls. However, even in the latter cases, the swelling of the starch granules upon cooking, due to the imbibition of water, causes the starch to become the predominant texture-affecting component.

Starch is a polymer of glucose with most of the bonds formed between the groups attached to the 1 and 4 carbon atoms. There is a linear polymer called amylose which has very few, if any, side chains (i.e., all of the bonds are 1,4) and a branched polymer with side chains of glucose molecules attached to the main sequence by 1,6 linkages. Fig. 9 illustrates these constructions (MacMasters and Wolff 1959). As a rule, starch granules contain both types of molecules, but waxy maize and waxy sorghum yield starches containing only amylopectin while starch from the wrinkled pea contains an unusually high proportion of amylose.

In addition to the glucose residues, starch molecules may contain small amounts of other groups. Potato starch contains appreciable proportions of orthophosphate present as end groups on the molecule.

Starches can vary not only in the proportions of amylose and amylopectin but also in the average molecular weights and the distribution of molecular weights in each of the two fractions. The viscosity of the linear chain fraction is a function of the molecular weight. For example, the intrinsic viscosity of the linear component of potato starch is greater than that of corn starch, while tapioca is intermediate, etc. The chain lengths are known to be in the same order. The viscosities of the branched fractions are not as simply related and are difficult to predict on

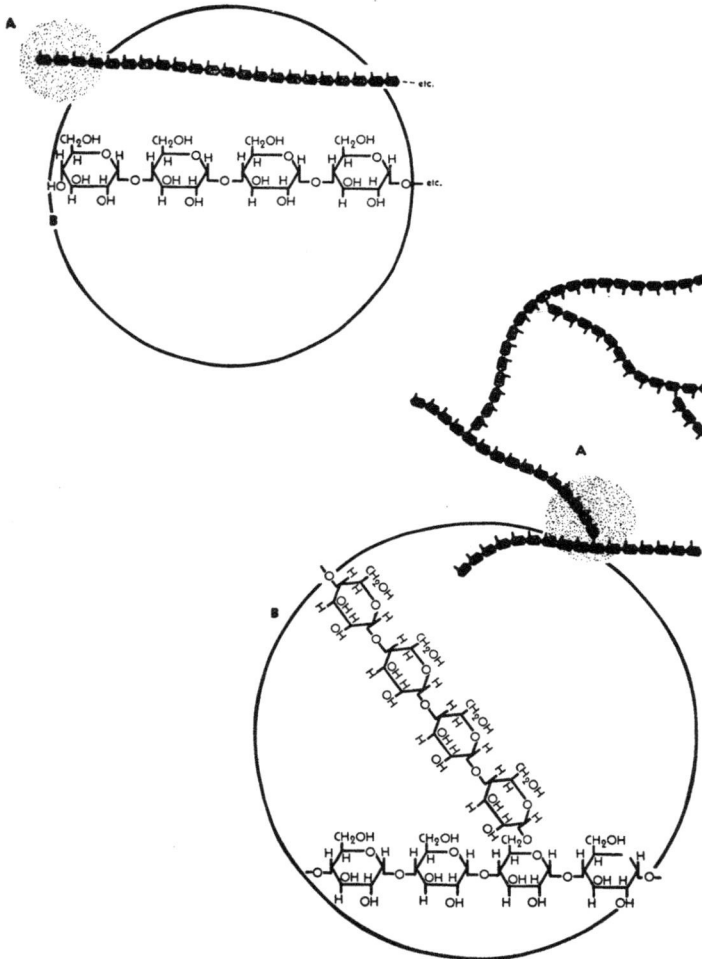

FIG. 9. STRUCTURE OF STARCH MOLECULES
Upper—amylose; Lower—amylopectin.

the basis of molecular weight. In general, gel strengths are also related to chain length and type of molecule.

Suspensions of native starches in water have no gel-like characteristics, regardless of the concentration. However, when the suspension is brought to a certain temperature, depending upon the source of the starch, the granules begin to swell (imbibe water) quite rapidly, eventually yielding a viscous paste or a gel whose rigidity depends upon the type of starch, its concentration, and the conditions of gelatinization.

When gelatinized starch pastes cool, viscosity increases and a firm gel may ultimately be formed, depending upon the kind and quantity of starch, the pH of the system, etc. Overcooking may adversely affect this property. Starches containing a high proportion of amylopectin (waxy starches) usually do not form rigid gels and yield instead soft pastes which are "long" in texture, i.e., they tend to stretch rather than break. Cornstarch pastes will set to a stiff "short" gel if at least five per cent of the starch is present. Wheat starch forms a softer gel at equivalent concentrations. Oat and rice starches set to soft "flowable" masses. The rheological behavior of some starch pastes during a heating and cooling cycle is illustrated by the Brabender curves reproduced in Fig. 10.

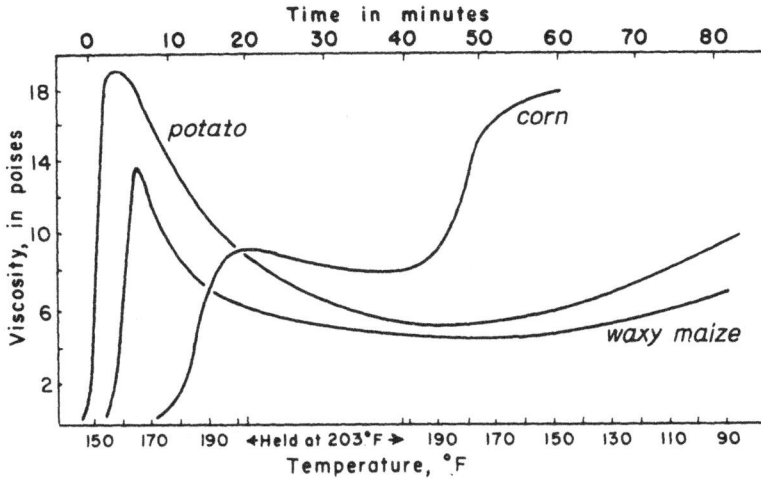

FIG. 10.   COOKING AND COOLING CURVES OF THREE STARCHES

An important phenomenon affecting the consistency of starch gels is retrogradation, the crystallization of the solvated molecules. The linear molecules which assumed a random orientation with respect to each other upon gelatinization become aligned and bound to each other by hydrogen bonds and van der Waals forces. Ultimately the gel is destroyed and the amylose precipitates as an insoluble floc. As shown by X-ray diffraction techniques, the precipitate consists of extended linear molecules situated in an orderly parallel array. It is impossible to redissolve the precipitate under any conditions practical for use with foods.

Amyloses from different sources show varying tendencies to retrogradation, corn and wheat starches being relatively labile while potato and tapioca starches form more stable gels. Amylopectin can be induced to retrograde under certain circumstances, but it is much more stable than amylose. Furthermore, retrograded amylopectin can be brought back to the gel state by moderate heating. The presence of amylopectin in native starches hinders the retrogradation of the amylose fraction.

If retrogradation is allowed to proceed slowly, as in a cooled dilute suspension, the molecules are able to align themselves parallel to other molecules along their entire length (or nearly so). In these cases the insoluble precipitate previously discussed is formed. When a hot concentrated cornstarch paste is rapidly cooled, the association is more rapid and less organized. Some parts of the molecules do not have an opportunity to align themselves properly, and an interlacing network of partially crystallized and partially amorphous polymers is formed. These processes are diagrammed in Fig. 11.

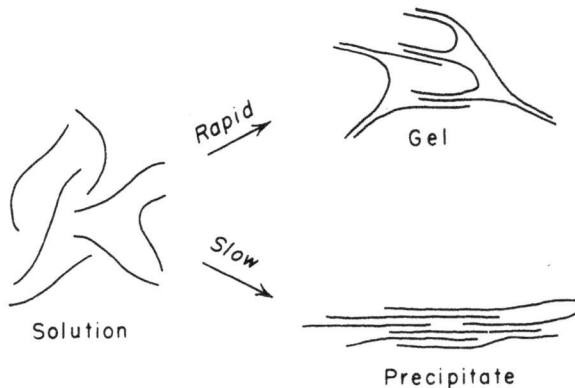

Courtesy Corn Industries Research Foundation

FIG. 11.   MECHANISMS OF STARCH RETROGRADATION

Starch probably has only a minor effect on texture in uncooked foods. In such products, most (if not all) of the starch is present in the form of granules of relatively low hydration. These granules are contained inside cells so that the perceived texture is due principally to the characteristics of whole cells and cell agglomerates. Breakage of cell walls during cutting, grinding, mashing, and other preparation procedures causes variable numbers of the granules to be released before cooking. In the cases of potatoes and rice especially, the quantity as well as the quality of the free starch is an important factor in determining the texture of the cooked product. In either case, an excessive amount of released granules creates an undesirable stickiness or pastiness in the food. This has long been known to be the principal cause of unpleasant texture effects in mashed potatoes, whether prepared from fresh, frozen, or dehydrated material.

Since the native starches which are available commercially differ so greatly in their pasting characteristics, it is frequently possible to select an unmodified starch to meet a particular texture requirement for a com-

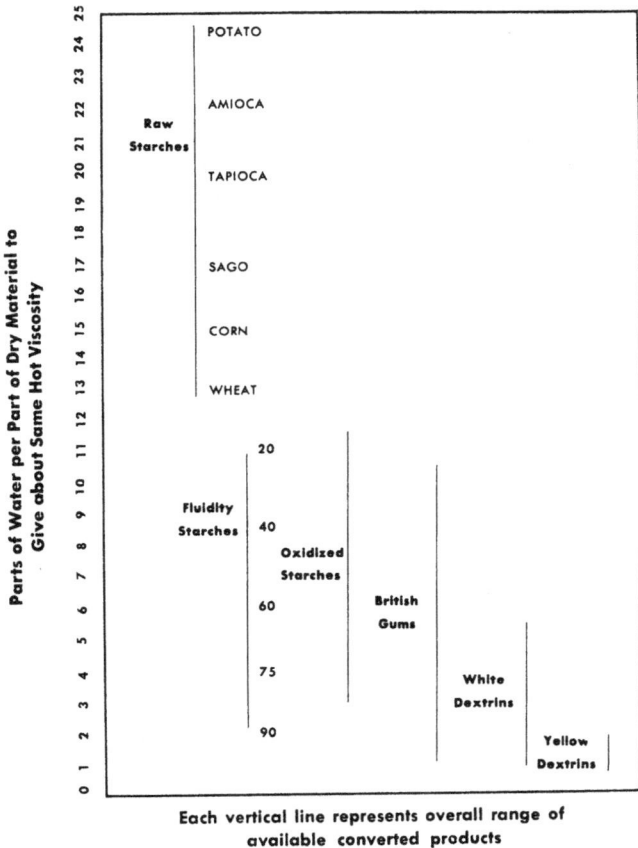

Each vertical line represents overall range of
available converted products

Courtesy Corn Industries Research Foundation

FIG. 12.   COMPARATIVE VISCOSITY RANGES OF VARIOUS NATIVE
AND MODIFIED STARCHES

pounded food product. In other cases, no native starch possesses the desired qualities or a suitable native starch exists but is too costly. Modified starches have been developed to meet many of these contingencies.

Most modified starches are derived from corn although some modified wheat and waxy cereal starches have been made available. Starches have been modified by pre-gelatinization, partial hydrolysis, oxidation, esterification, etherification, or other physical or chemical alterations. Fig. 12 illustrates the relative viscosities of some common modified and native starches.

Pre-gelatinized starches are prepared by heating aqueous suspensions of the material until gelatinization is complete and then roller-drying the gel. These products will form viscous pastes or gels when mixed with cold or warm water and thus are suitable for use in instant puddings and pie-fillings.

Acid-modified starches are prepared by heating a starch slurried with dilute sulfuric or hydrochloric acid to about 120°F. for several hours. The chemical effect of this treatment is a reduction in the chain length of the polymer molecule. The granules also become more fragile. From a practical standpoint, the important effects are a decrease in the viscosity of the hot pastes and an increase in the rigidity of the cold gels. The former change increases the ease with which concentrated suspensions can be handled during processing. Acid-modified starches are frequently used in gum drops, jelly beans, and other confections requiring a very firm gel.

The use of esterified or etherified starches in the food industry has been held back somewhat by toxicity considerations. Doubtless many varieties of these modified starches will be shown to be innocuous and will find use in foods.

Pectin Gels

Pectins are linear polymers of galacturonic acid in which the carboxyl groups are partially esterified with methanol. Fig. 13 is a diagram of a pectin molecule. Molecular weights as high as 200,000 may be found. Pectins are always found in association with arabans, galactans, and other polysaccharides. The consensus of workers in the field is that the latter compounds are present as contaminants rather than as essential or bound parts of the pectin molecule.

Pectic compounds are doubtless present in all fruits and vegetables. In many of these cases they make important contributions to texture. Traces have been found in cereals, where they have little effect on texture. Pectic substances are found in cells as cementing substance in the middle lamellae and as thickenings on the cell wall. The latter material can be solubilized readily by heating with dilute acid. Pectic substances of the middle lamellae remain insoluble in hot dilute acid but are brought into solution by weak alkalies or ammonium oxalate. The native substances which yield pectin when extracted by acid or other means are called protopectin.

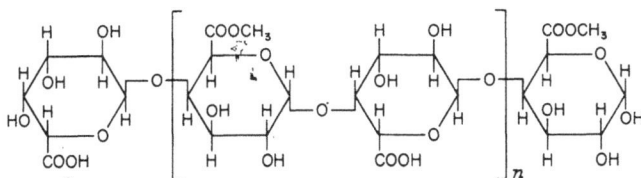

FIG. 13. POSSIBLE STRUCTURE OF A PECTIN MOLECULE

Pectin disperses in water to form a viscous colloidal sol. A very important property, so far as food texture is concerned, is its ability to form sugar-acid-pectin gels, i.e., the fruit jellies, jams, and preserves of commerce. Gel formation of pectin with sugar and acid was shown by Olsen (1934) to be in accord with the following assumptions:

(1) Pectin is a negative hydrophilic colloid.

(2) The sugar functions as a dehydrating agent.

(3) Hydrogen ions function by reducing the negative charge on the pectin, thus permitting coalescence of the molecules to form a network.

(4) Dehydration of the pectin requires time to come to an equilibrium, so that some gels set more quickly than others or require different setting temperatures.

(5) The rate of dehydration and precipitation increases directly as the hydrogen ion concentration increases.

(6) Maximum jelly strength is reached at equilibrium.

The preceding statements are still valid although they have been challenged from time to time.

The nature of the cross-bonding between pectin molecules which is responsible for the rigidity of their gels has been explained as being due to: (1) carboxyl groups, (2) ester groups, or (3) "dehydration and electric discharge induced by the sugar and H-ions." This problem has not been fully resolved, but the bulk of the evidence appears to favor the ester linkage as the responsible group. Owens *et al.* (1954) state

. . . although van der Waals and the ionic forces due to resonance of ester groups are weak, there are so many groups available that in summation they could be centers of crystallization. Further cooling of the gels enhances hydrogen bonding between the various polar groups, which could provide sufficient bonding strength to account for the rigid, strong gels that can be made even with very small concentrations of pectin.

They regard "bridging" by sucrose molecules to be of minor importance.

Cheftel and Mocquard (1947) investigated the rheological properties of pectin gels of the high sugar-high methoxyl type. When these gels were subjected to large deformation, they showed no internal strains under crossed Nicols. Total breaking strength was always reached even with the smallest forces which could be obtained with the apparatus used by these investigators. and typical plastic flow occurred under light loads. From these observations, it was concluded that all previous data on elasticity or rigidity of pectin gels were in error. (These conclusions were not applied to low methoxyl gels, which do possess elasticity.)

The conclusions of Cheftel and Mocquard were disputed by others, for example Owens *et al.* (1954) who point out that deformation of high methoxyl-high sugar pectin gels is linear with respect to small stress and recovery is complete. This linearity and recoverability has been proved with torsion testers and plunger devices when the deformation is kept small. Hooke's law is therefore valid and pectin gels are elastic provided the stresses are small and are applied for reasonably short periods.

The setting time, or the time required for a pectin jelly to assume a certain rigidity under controlled temperature conditions, is an important factor in commercial use. Hinton (1950) points out that the setting temperature of a jelly is a determinate physical property, whereas the setting time is a secondary property dependent upon the temperature conditions. The time of setting for a certain jelly batch is governed by the velocity at which the setting temperature of the mixture is reached. Doesburg and Grevers (1960) made an intensive study of factors affecting setting time. They found that the setting time of jellies containing pectins with a low ash content increases with decreasing degree of esterification of the pectin down to about 50 per cent, while jellies from pectins with lower degrees of esterification exhibit shorter setting times. The setting time of jelly batches is shorter at lower pH. Higher sugar contents decrease setting time. The jellying power of partially saponified pectins is considerably above the jellying power of pectins with high degrees of esterification from which they were made.

When the methyl ester content of pectin is reduced below about seven per cent, the substance loses its ability to form the sugar-acid gel and gains the ability to form a gel with calcium. This behavior is almost certainly due to the formation of bridges between carboxyl groups on adjacent pectin molecules by the divalent calcium ion. Rapidly gelling dessert powders can be made from demethoxylated pectin and a slowly dissolving source of calcium ion. These systems are also useful for low calorie desserts since the high concentration of sugar necessary for pectin gels can be reduced or even omitted entirely with the use of non-caloric sweeteners. It appears that these gels have sensory properties different from sugar-acid-pectin jellies and that it is difficult to correlate these differences with measurements of the rheological properties of the gels.

Vegetable Gums and Mucilages

This group of substances is composed of exudates formed on the bark of certain species of woody plants and products obtained by aqueous or alkaline extraction of seeds or of the structural framework of algae, etc. Some of these materials will form gels under certain conditions while others are not capable of forming elastic gels under any ordinary circumstances and so, strictly speaking, are not within the purview of this chapter.   In most cases it appears that the basic element of the gum or mucilage is a complex polysaccharide acid.   Among the important vegetable gums are agar, algin, carrageenin, tragacanth, gum arabic, gum guar, and gum karaya.

Agar is the dried extract of *Gelidium, Gracilaria,* and related red algae. The active gel-forming compound is probably a linear polygalacto-pyranose sulfuric ester.   Some authorities (Selby 1954) do not regard the sulfur as an essential constituent.   The free acid will not gel, but it occurs in nature in combination with calcium, and, when this ion is present the polymer will form very firm gels at relatively low concentrations.   The molecular weight of the substance is quite large.

Agar is insoluble in cold water, slowly soluble in hot water, and readily soluble in boiling water.   At the higher temperatures it forms sols of rather low viscosity and the viscosity is only slightly affected by temperature until the gelling level is reached.   Agar is neutral and its gelling power is affected by pH, reaching a maximum between pH 8 and pH 9.   Electrolytes reduce the swelling power of agar, apparently by competing for the sites binding calcium.   Alcohol and other organic solvents reduce the gelling efficiency, probably by dehydrating the polymer molecules; although some part of the action may be due to a reduction in the net electric charge of the molecule.

Agar gels show a high degree of temperature hysteresis.   A gel formed by dissolving agar at about 200°F. will not solidify until the temperature is reduced to about 100°F.   The gel will then retain its form when heated to intermediate temperatures.   The gel strength is approximately proportional to the amount of agar between 0.5 and 2.0 per cent.   On aging, agar gels of low concentration are subject to syneresis and weeping. Agar is used in confectionary such as Turkish pastes, as a stabilizer in sherbets, cake icings, and custards, etc.   A good discussion of the rheology of agar has been published by Cheftel and Mocquard (1947).

Irish moss or carrageenin is prepared from the perennial red algae *Chondrus crispus* or *Gigartina mamillosa.*   The substance responsible for the gelling properties is a mixed salt of the sulfate ester of a polysaccharide complex composed principally of D-galactopyranose with some L-galactose units, 2-ketogluconic acid units, and non-reducing sugar units, each

TABLE 6

CHARACTERISTICS OF SOME NATURAL FOOD GUMS AND MUCILAGES

Common Name	Obtained From	Chemical Nature	Principal Residues	Typical Viscosity[1]
Agar	Algae (seaweed)	Mixture of polysaccharides	D-galactose, sulfate, 3,6-anhydro-L-galactose	Gels
Algin	Brown algae (seaweed)	Complex uronic acid polymer	D-mannuronic acid, L-glucuronic acid	1,800
Carrageenin	Red algae (seaweed)	Polysaccharide sulfate ester	D-galactose, sulfate, 3,6-anhydro-D-galactose	225
Guar gum	Seeds of a legume	Polyhexose	D-mannose, D-galactose	3,000
Gum arabic	Exudate of a tree	Calcium, magnesium, and potassium salts of arabic acid	D-galactose, L-arabinose, L-rhamnose, D-glucuronic acid	Low
Gum tragacanth	Exudate of a shrub	Composite of a complex acid polysaccharide and a neutral araban	L-arabinose, D-xylose, L-fucose, D-galactose	3,200
Karaya gum	Exudate of a tree	Complex acid polysaccharide	D-galacturonic acid, L-rhamnose, D-galactose	2,300
Locust bean gum	Seeds of a tree	Galactomannan	D-galactose, D-mannose	2,750

[1] Viscosity in centipoises of a heated and cooled one per cent solution measured at 77°F.

TABLE 7

SOME REACTIONS OF FOOD GUMS AND MUCILAGES

Common Name	pH[1]	Gel Former?	Effect of HCl	Response of Sols to Different Agents		
				Effect of NaOH	Effect of Salts	Effect of Heat
Agar	7	Yes	Viscosity decreased	Increases viscosity to pH 8.5, thereafter decreases viscosity	Effects are slight	Solid gel to 200°F.
Carrageenin	7	Yes	Viscosity decreased	Viscosity decreased	Gelling facilitated	Gels melt near 100°F.
Guar gum	7	No	Effects are slight	Effects are slight	Borates gel	Viscosity decreased
Gum arabic	5	No	Viscosity decreased	Viscosity increased to pH 7	Borates gel	Viscosity decreased
Gum karaya	4.6	No	Viscosity decreased	Viscosity increased	Viscosity decreased	Viscosity decreased
Gum tragacanth	5.5	Yes	Viscosity decreased	Viscosity increased to pH 8, thereafter decreased	Effects are slight	Viscosity decreased
Locust bean gum	5.3	No	Viscosity decreased	Low concentrations decrease viscosity, high levels cause an increase	Borates gel	Viscosity increased up to 180°F.

[1] pH of a fresh one per cent solution.

of the latter being the size of hexose and combined with one sulfate radical (Stoloff 1954). The so-called lambda form of carrageenin has anhydro-D-galactose units in the chain in addition to the D-galactose residues. The kappa form has no anhydro units. Sodium, potassium, calcium, and magnesium are the cations usually found in Irish moss extracts.

The properties of carrageenin are similar to those of agar except that concentrations of about three per cent are required to yield firm gels. The viscosity of a one per cent sol at 75°F. is about 150 cps. At low concentrations the addition of salts lowers the viscosity of sols progressively until a constant value is reached, but the strength of the gel and its rigidity are increased. Carrageenin forms a complex with some of the milk proteins and so is useful as a stabilizer in chocolate milk. It has been used for its texture enhancing effects in whipped cream, soups, and ice cream.

Gum arabic, gum karaya, gum guar, and locust bean gums do not form gels under normal conditions of usage. The first two are prepared from tree exudates while the last two are manufactured from seed pods. Gels can be formed by adding borate ion to the sols, but this technique is obviously inapplicable to foods.

Gum tragacanth will form thick gels in concentrations of about two per cent. It swells in water to give highly viscous solutions, the viscosity of a one per cent solution of the highest grade being about 3,400 cps. Viscosity of solutions tends to increase upon aging up to about 24 hours. Maximum viscosity is reached at pH 8 with a rapid decrease occurring on either side of this optimum, but there is apparently no direct relationship between pH and gel-forming ability. Gum tragacanth can be used in confectionery jellies and allied products.

Tables 6 and 7 summarize some of the important properties of the vegetable gums and mucilages.

Cellulose Gum

Native cellulose fibers have no gel-like characteristics. However, pure (99+ per cent glucose residues) crystalline cellulose in colloidal size particles (average width 50 microns) has been shown to form gels when dispersed in water under the proper conditions. Preferred concentrations are around 15 per cent. These gels are sensitive to the amount and kinds of ions found in hard water and are best made with distilled water. They appear to be stabilized or improved in texture by addition of small amounts of vegetable gums. Colloidal crystalline cellulose could be useful as the texturizing ingredient in butter-flavored spreads for use in all climates, and in dietetic cake icings, pastry fillings, fudge-like confections, mayonnaise, etc. Gels made with this substance do not appear to show true elastic response although thorough rheological studies have not been made.

Cellulose modified by the introduction of carboxymethyl groups on to the chain has pronounced swelling properties and can be used to form gels. Generally, gel formation requires the addition of the salt of some polyvalent cation. Basic aluminum acetate has been used for this purpose. In foodstuffs, carboxymethylcellulose has been used more as a stabilizer, emulsifier, and viscosity improver than as a gel former.

Gelatin

Gelatin is a polymer of amino acids. It is prepared from collagen, the principal intercellular constituent of the white connective tissue of animal skins and bones, by alkaline or acid hydrolysis followed by extraction with hot water. It is not a well-defined substance chemically, being a collection of molecules of varying composition and size. In many samples of gelatin, about half of the molecules will have molecular weights in the range of 10,000 to 60,000. Gelatins degraded to an average molecular weight of about 10,000 to 15,000 will not form gels at 32°F.

Gelatin differs little in chemical composition from collagen. The hydrolytic processes used in its manufacture separate the polypeptide chains in collagen and break them into shorter segments along their length. Acid- and alkali-processed gelatins are considerably different in their chemical and mechanical properties due, evidently, to variations in points of scission and to different effects of the reagents on the side groups attached to the amino acid moieties.

A distinguishing feature of gelatin is the formation of rigid gels from cooled aqueous sols of rather dilute concentration, even below one-half of one per cent. As stated in the excellent review of Idson and Braswell (1957) the gel-sol transformation is responsive to temperature approximately according to:

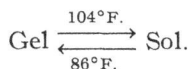

$$\text{Gel} \underset{86°F.}{\overset{104°F.}{\rightleftharpoons}} \text{Sol.}$$

Gelation occurs as the result of the binding together of individual molecular chains by secondary attractive forces localized at widely separate points. The locus of attraction may include several amino acid residues. Hydrogen-bonding is probably the principally effective force joining the molecules.

The rheology of gelatin has been reviewed recently by Ward and Saunders (1956). The effect of changes in rate of shear on the apparent viscosity of gelatin is small in dilute solutions. In more concentrated solutions, above the gelation temperature, the behavior is not fundamentally different, apart from the usual hydrodynamic action. Viscosity increases approximately in an exponential relationship with increases in gelatin concentration. The viscosity increases approximately exponentially with decreases in temperature.

Gels of gelatin obey Hooke's law. According to Idson and Braswell, methods used for measuring the rigidity modulus of gelatin gels usually involve: (1) bending or stretching gel strips, (2) compressing gel cylinders, (3) subjecting rods to torsion by rotating one end, (4) measuring the torque exerted by gelatin in the annular space between a rotating cylinder and a cylinder free to move against a restoring force, (5) observing the distortion of the meniscus of a gel under air pressure, and (6) determining the velocity of propagation of transverse vibrations in a gel sample.

Rigidity of gels increases about as the square of the concentration and in inverse relation to the temperature. In the middle temperature and concentration ranges, at least, the plot of temperature versus modulus of rigidity divided by the square of the concentration is linear. Rigidity is rather insensitive to variations in hydrogen ion concentration in the region near neutrality. After setting, the rigidity of gels increases with time, rapidly at first and then more slowly, the rate approaching an asymptote.

If loading times are kept short, i.e., a few seconds, the stress required to maintain a constant strain on a deformed gel remains constant. If the stress is maintained for relatively long periods, "creep" occurs and the gel is permanently deformed. Workers in this field have postulated that the stress decay may be due either to breaking and re-establishment of cross-linkages or to continuing growth of oriented crystallites.

The principal consumption of gelatin is in dessert and salad bases. It is also used in some luncheon meats and to congeal the juices in canned hams. Its gel-forming properties are important in several kinds of confectionery, particularly marshmallows.

Egg Albumen

Normal egg white contains about twelve per cent solids and about 85 per cent of these are proteins. In unmodified form the proteins form a weak gel. Denaturation of these proteins by heat yields the familiar firm gel of cooked egg white. The use of egg white in the food industry is based on its ability to form foams rigid enough to support relatively large amounts of other ingredients.

The white of a hen's egg is quite heterogeneous. Major divisions are an outer and inner thin white and an intermediate thick white. The proportion of thick white is the principal determinant of the appearance of poached and fried eggs. The thin white is practically devoid of elastic properties, but the thick white responds elastically to measurements as though it were a weak gel containing a system of microscopic elastic fibers. Apparently fibers of insoluble keratin-type protein called ovomucin do exist in the native product.

Egg white coagulates very rapidly at 130° to 135°F. and is almost instantaneously denatured at 140°F. High concentrations of sucrose, some inorganic salts, and high protein concentrations tend to elevate the coagulation temperature. Forsythe (1960) indicates that coagulation or precipitation is not observed in the absence of salts. The phenomenon is also retarded by high pH.

A complete discussion of the rheological properties of egg white can be found in the fine review of Forsythe (1960) and in an article by Brooks (1960).

GELS AND TEXTURE IN NATURAL FOODS

By natural foods we mean those foods which have had a minimum of processing before they were cooked. They are products which have not been compounded from two or more ingredients during processing. The influence of gels on texture is most obvious in manufactured foods such as starch puddings and gelatin desserts, but many natural foods consisting principally of agglomerates of whole cells also exhibit important textural phenomena due to gel structures.

The composite of proteins and other compounds constituting the protoplasm of intact cells is a gel, of course, but it does not have a direct and important effect on the perceived texture of foods consumed as pieces containing many whole cells. In these foods, the major texture effects are due to cell turgidity, cell wall characteristics, and cohesive forces between the cells. Caviar may owe a considerable part of its textural properties to protoplasm.

Cooked foods containing starch often have their texture influenced by this substance because the swelling which precedes and accompanies gelatinization profoundly affects the texture of the whole cells. Pectin-containing foods not only leach pectic-substances into the cooking water when they are boiled, but may also have their intercellular adhesion altered by these conditions so that even large pieces owe an important part of their texture to the gel.

Potatoes

The desired texture of cooked potatoes is usually described as "meal-iness," that is, they should have a dry appearing structure which readily crumbles into fragments when pressed with a fork or other implement. The crumbly texture as opposed to a more cohesive structure is obviously related to the degree of intercellular adhesion and therefore to the properties of the pectinous substances in the middle lamellae. These properties will be described in detail in Chapter 6.

It is important that free starch be at a minimum in cooked potato products since an excess leads to a sticky gelatinous texture which is undesirable in any type of potato dish. The amount of free starch found in potato preparations is related to the number of ruptured cells, and Hall and Fryer (1953) recommended direct counting of cells as a means of predicting the texture quality of dehydrated potatoes. Reeve and Notter (1959) also found a close agreement between the extent of cell rupture and stickiness of mashed potatoes prepared from dehydrated material. They indicate that stickiness was pronounced in samples with ten per cent or more of ruptured cells.

The pasting characteristics of starch from different varieties of potatoes seem to vary sufficiently to affect the perceived texture of the cooked material. Potato starch gels as a group tend to be extensible and sticky. They may not show a sharp breaking point with the Saare disc apparatus or the Fuchs gelometer (Hofstee and de Willigen 1953). The instantaneous elastic deformation which follows the application of force to these gels is followed by "creep" or permanent deformation. When the gel is unloaded, an instantaneous partial recovery occurs, followed by a retarded recovery phase.

Unrau and Nylund (1957) found that the maximum relative viscosities of heated suspensions of lyophilized potatoes were related to the subjectively evaluated mealiness. These viscosity results were obtained with the Brabender Amylograph and are clearly associated with variations in the amount and pasting characteristics of the starch. Kuhn et al. (1959) also reported data bearing on this point. They found that the changes observed in the relative viscosity during heating of potato slurries were different for each variety. Mealier potato varieties tended to gelatinize more rapidly and to attain a higher maximum relative viscosity. Specific gravity of the uncooked material and the per cent moisture gain during cooking were associated with the degree of mealiness as evaluated by panels. The significance of these observations with respect to the starch content of the potato was not discussed by Kuhn et al., but it is certainly possible that the quantity and quality of starch affects both the specific gravity of the raw potato and the absorption of water during cooking.

Rice

Although different cultural groups may prefer variant kinds of texture, there is a rather widespread preference throughout the world for rice that is tender but not mushy or sticky when cooked. Each variety has a typical cooking quality which does not vary much from lot to lot. Most of the long grain varieties tend to cook to a dry and fluffy state and the kernels do not split or stick together. Short grain varieties are usually not as soft when fully cooked and tend to stick together more than the other types (Beachell 1959). These organoleptic qualities have long been recognized as a function of the gel characteristics of the starch component of the grain.

Amylose proportion and gelatinization temperature are factors affecting stickiness. Rao *et al.* (1952) found a correlation between amylose content and the swelling number of rice (the latter figure is closely associated with texture according to a paper on the physico-chemical aspects of the curing of potatoes in order to improve storage and cooking quality by Rao [1948]).

The gelatinization curves of suspensions of ground starch as obtained with a Brabender Amylograph or Viscograph are useful in evaluating rice properties. Short grain varieties tend to gelatinize in the lower temperature range, and there appears to be a general correlation (negative) of temperature with stickiness. There seems to be little doubt that the major determinant of rice texture is the starch, even though cell cohesiveness must also play a role.

Newly harvested rice tends to leach more of the cell contents into the cooking water than does rice which has been stored for some time (Desikachar and Subrahmanyan 1959). As a result, fresh rice tends to cook to a pasty mass enclosing a viscous sticky gruel. Desikachar and Subrahmanyan attribute these differences to a qualitative difference between the cell walls of new rice and old rice. However, the data of these authors do not preclude the possibility that the cell contents undergo changes during aging which inhibits their diffusion from the cell when the membrane of the wall is damaged by cooking. Such changes could result from alterations in the granule organization of the starch or from denaturation of the protein, the latter causing interference with diffusion through relatively large discontinuities in the membrane. Furthermore, non-enzymatic browning reactions (which are possible under the conditions existing in the cells) could result in the formation of insoluble materials.

Alimentary Pastes

Macaroni, spaghetti, and the like are prepared by extruding a cold dough made from durum semolina and water through an orifice. This dough is dried slowly under mild temperature conditions until it reaches a moisture content approaching that of the dry semolina. Changes due to enzymes from the grain and to microbiological activity undoubtedly occur during processing but they evidently have little effect on the rheological properties of the food.

One of the most extensive investigations of the rheological properties of *cooked* alimentary paste was reported by Shimizu *et al.* (1958). Typical noodles from Japanese domestic wheat flour exhibited instantaneous elasticity of the order of 10^5 dynes per sq. cm., a viscous flow of about 10^8 poises, and a retarded elasticity under constant load. The elastic modulus decreased with increases in temperature, indicating that both entropy and internal energy changes occurred when noodles were deformed. The elastic modulus, breaking energy, and stress relaxation increased with increasing crude protein content and boiling times. These investigators concluded that the gluten network played an important role in the gelatinized and denatured dough but that the predominant influence on rheology was due to starch.

Harris and Sibbitt (1958) found that macaroni toughness, as indicated by a panel tenderness score, increased with protein content. Location, year of growth (weather conditions), and the durum variety also have effects on tenderness.

MEASUREMENT OF GEL RHEOLOGICAL PROPERTIES

The value of the gel-formers used in foods is a function of the strength of the gel formed by a given concentration of the substance in water, or, perhaps more accurately, of the minimum concentration required to give a gel of the desired rigidity. Since the economic importance of this factor is obvious, attempts were made early in the development of the materials to secure meaningful objective measurements of it. However, the rheology was almost always poorly understood and no correlation between rheological properties and sensory qualities was even attempted. Consequently, the testing methods were often based on unsound premises, commonly taking the form of a procedure intended to duplicate as closely as possible the actual commercial use of the material. In the evolution of these procedures, it was frequently found that the results obtained on the same sample by different laboratories did not agree as closely as was desired. The differences were attributed to unrecorded variations in procedures, materials, or equipment. As a result, the specifications for these factors were made more and more detailed. The pectin jelly test of the Pectin Standardization Committee of the Institute of Food Technologists is an interesting example of the end product of an evolution of this type. Since this procedure also represents the current "standard" test for pectin grade in this country, it is reproduced in full.

(A) **Preparation of Jelly.**—Calculate the weight of pectin to use by dividing 650.0 by the value of an assumed firmness grade (a) for the pectin.

(a) Jellies produced by this method should contain 650.0 gm. of total soluble solids in 1000.0 gm. of jelly. The ratio of the actual weight of *total soluble solids* in a *particular* jelly to the weight of the pectin in *that* jelly is defined as the assumed grade for the pectin, in that specific jelly. Hence, the weight of pectin to use in a jelly is given by the expression: 650/assumed grade.

Weigh into a dry container the amount of sugar needed which will be 650.0 minus the weight (to the nearest gram) of pectin used (b).

(b) Examples.—Suppose one assumed a "firmness grade" of 150, then 650/150 or 4.33 gm. would be the weight of pectin to use and 650–4 or 646 would be the grams of sugar to use. The sugar should be the finely granulated type made for table use, where about 75 per cent is within the band of 35 to 80 mesh.

Transfer about 20 to 30 gm. of the weighed sugar into a dry 150 ml. beaker and add the weighed pectin sample. Mix the pectin and sugar thoroughly in the small beaker by stirring with a spatula or glass rod.

Put 410 ml. of distilled water into a 3-qt. stainless steel sauce pan (c) containing a stainless steel potato masher (d) for stirring. (The sauce pan and stirrer should have been tared previously, on a scale or balance.)

(c) Ekcoware 3-qt. stainless steel pan No. 7323, Flintware 3-qt. pan No. 7623 or Revere 3-qt. copper-clad pan No. 1403 are excellent examples of pans suitable for this purpose.

(d) A Flint stainless steel potato masher No. 1905 made by Ekco Products Company, Chicago 39, is ideal for stirring and skimming pectin jelly test batches. The pectin-sugar mixture is now poured into the water all at once (e) then gentle stirring is started and continued for about 2 minutes.

(e) The object here is to get the sugar-pectin mixture *under* the surface of the water as quickly as possible. When hot water is used the sugar dissolves too quickly and the pectin tends to stay on the surface and stick to the sides of the pan. Avoid splashing when stirring so pectin is not scattered to upper parts of pan. Just before the pan is put on the heater or stove any traces of pectin-sugar remaining in the small beaker should be transferred to the jelly batch.

The jelly kettle or sauce pan is then placed on a stove and heated until the contents come to a full rolling boil, stirring being continued during this period. The remaining sugar is added and heating and stirring continued until the sugar is dissolved. Heating is continued until the net weight of the jelly batch is 1015.0 gm. (f).

(f) The gas or electric stove should be adjusted *so that the entire heating time for the jelly is 5 to 8 minutes!* When an electric stove is used it should be preheated. The stirrer is to be left in the kettle during the cooking and weighing period. If the batch weighs less than 1,015 gm., distilled water is added in slight excess so that additional boiling will be necessary to reduce the net weight to 1,015 gm. The amount of water used at the start should be adjusted by each operator so that the entire heating time is kept within the 5 to 8 minute range.

After the 1,015 gm. batch is removed from the balance or scale, it is allowed to sit undisturbed on the desk top for one minute, then the pan is tipped so that the contents are nearly ready to overflow, and any foam or scum is quickly skimmed off. The stirrer is removed and a thermometer is put into the jelly batch. While the kettle is tipped the batch is stirred gently with the thermometer until the temperature reaches exactly 95°C. The batch is then *poured quickly* into three previously prepared Ridgelimeter glasses, each containing 2.0 ml. of tartaric acid solution (g).

(g) Acid solution.—The acid solution should be made by dissolving 48.80 gm. of tartaric acid crystals in distilled water and making up to a total volume of 100.0 ml. in a volumetric flask.

Preparation of Glasses.—The Ridgelimeter glasses are Hazel-Atlas No. 85 tumblers which have been ground down so that the inside height is exactly 3.125 inches. Scotch drafting tape or Mystik masking tape, $3/4$ inch wide and $9^1/_2$ inches long is used to make sideboards on each glass. The strip should cover the top $1/_4$ inch of the glass and must extend $1/_2$ inch *above* the glass. Squeezing the tape against the glass, especially under the flange near the top of the glass, will insure a tight seal which will not leak hot jelly.

Cooling to 95°C.—Pectins whose jellies set at different rates are difficult to grade precisely unless their jellies are poured at some very definite and constant temperature. This fact was recognized by Stuewer, Beach, and Olsen (1934) when they said, "Observations made here indicate that most of the difficulties encountered in the routine evaluation of apple pectin jellies may be obviated by the simple expedient of preparing the jelly mixture without acid, adding an excess of acid to the glasses, and pouring the jelly mixture into these at exactly 96°C."

It is known that the acid concentration (by titration and by pH) as well as the soluble solids can differ from top to bottom in a glass of jelly, even when the acid is added to the batch before it is poured. The differences in acidity are too small to be significant when the pH is below about 3.0.

Pouring the Jelly.—A glass rod should be in only one of the jelly glasses. When the jelly batch has cooled to 95°C. it is poured in the first glass as rapidly as is consistent with reasonably accurate filling, the jelly being stirred vigorously with the glass rod *only* during filling, *but no longer.* The glass rod is put into the second glass, pouring it as before, and so on for the three glasses. It is best to pour very rapidly until the glass is filled part way up the sideboards, then to pour more slowly so that the glass can be filled completely full to the point of overflowing. There is ample evidence to show that the 2.0 ml. of acid gets suitably mixed in the glass of jelly when rapid pouring is done as mentioned above.

About 15 minutes after the glasses are filled they are covered with their regular metal lids which can be fitted snugly over the 'sideboards.' The jellies should be stored 20 to 24 hours, at 25°C. ± 3.8°C. (h).

(h) If room temperature is outside the range of 22 to 28°C., allow the jellies to cool to about 30° to 35°C. then put them into an incubator at 25° ± 3°C. for 20 to 24 hours.

(B) **Measurements on the Test Jelly.**—(1) *Determination of the Jelly Sag.*—After the jellies have been stored for 20 to 24 hours, the lids are removed from the glasses and tape strips torn off. A tightly stretched wire, clean and wetted (cheese cutter furnished with the Ridgelimeter) is carefully drawn across the top of the glass while the latter is held upright and is turned slowly part way around, so that a smooth cut is made to remove the layer of jelly projecting above the top of the glass. The detached top layer is carefully removed and discarded.

The jelly is turned out of the glass into an inverted position on a plate glass square furnished with the Ridgelimeter. This is accomplished by holding the glass tilted at about a 45° angle while the point of a spatula is inserted between the top of the jelly and the glass, to start the separation of the jelly from the glass. The jelly should pull away from the glass while the latter is rotated slowly, without further aid from a spatula. The glass is quickly and carefully inverted just above the glass square in such a way that the jelly slides out and stands upright near the center of the glass plate. Do not drop the jelly onto the plate!

A stopwatch is started as soon as the jelly is on the glass plate. If the jelly leans slightly to one side, this usually can be corrected by gently tilting the glass plate away from the direction in which the jelly leans. The plate and jelly should now be placed carefully on the base of the Ridgelimeter so that the jelly is centered under the micrometer screw, which then should be screwed down *near* to the surface of the jelly. (The Ridgelimeter should be used only on a *level* desk or table.)

Two minutes after the stopwatch was started, the point of the micrometer screw is brought just into contact with the top jelly surface. Illumination should be arranged so that contact of the micrometer tip with the jelly surface can be observed easily. The lowest line on the vertical scale beyond which the lower edge of the circular micrometer head has passed, is the per cent and the number on the micrometer head nearest the vertical scale denotes the tenth of a per cent sag. The Ridgelimeter reading is recorded only to the nearest 0.1 (j).

(j) Check of the Ridgelimeter calibration.—A $^3/_8$ inch brass rod exactly 2.500 inches in length may be used to check the Ridgelimeter scale and *must* be used to adapt the instrument for using new glass plates. A slight shift in the vernier and/or in the scale, or use of a glass plate other than the one originally sent out with the instrument, can result in unreliable readings. It is easy to check the instrument by standing the 2.500 inch gauge rod upright on the glass square, under the tip of the micrometer screw. When the tip of the screw is in *gentle* contact with the gauge rod, the instrument should read exactly 20.0. Both the vertical scale and the vernier knob can be reset so the instrument *does* read exactly 20.0 when checked with the 2.500 inch rod as just described. Occasional checking by this method is recommended to insure that the scale settings remain fixed.

When Ridgelimeter readings on different glasses from the same jelly batch differ more than 0.6, the batch should be remade.

(2) *Determination of Soluble Solids and pH of Jelly.*—After the second glass poured has been tested on the Ridgelimeter, cut it into halves lengthwise and remove a small portion from the center of the jelly and spread it quickly over one of the refractometer prisms. Quickly close the refractometer, firmly, and

after about a minute or so, read the instrument, recording also the temperature of the prisms (the water circulating through the instrument) so that later
corrections can be made for getting the soluble solids at 20°C. by the table of
corrections in the appendix of *Methods of Analysis* (k) 8th Edition, page 886,
Assoc. Offic. Agr. Chemists, Wash., D. C., 1955.

(k) When the soluble solids content of these test jellies is as much as 1.0
unit from 65.0 per cent at 20°C., serious errors must have entered into the
method of making the jellies. A spread of ±1.0 per cent soluble solids can
mean an error of ±3 to 4 per cent in jelly grade, a larger error than is involved
with ordinary use of the Ridgelimeter. It is necessary, therefore, to get correct
soluble solids readings by being careful that samples for testing are always
taken quickly from an unexposed jelly surface, and that temperature is given
due consideration.

Mention was made in Note (g) that the acidity and soluble solids may vary
from top to bottom of a glass of jelly, regardless of whether or not acid was in
the batch or in the glass. These variations are too small to be of consequence
when the pH is considerably below 3.0, as in the present method, and they vary
in magnitude and even in direction, from one pectin type to another.

The pH can be determined by taking a portion of jelly from the center of
the second glass and crushing it into a small beaker where a glass electrode
system may be inserted into the jelly mass. If the pH equipment is in a suitable
condition a value accurate to a few hundredths of a pH will be attained in a
few moments.

The jelly test of the Pectin Standardization Committees is an example
of a "sag" test, several of which are used in evaluating gel-formers.
Christensen (1954) classified tests made on pectin-acid-sugar jellies into
(1) those tests in which the elastic limits (breaking strength) of the
jellies are exceeded and the jelly is ruptured, and (2) those tests measuring deformation (sag) of jellies without exceeding the elastic limits.
Into the first category he places such tests as Sucharipa's jelly disc method,
the Fellers-Clague penetrometer, the Lüers-Lochmüller Pektinometer,
and the Delaware jelly strength tester of Tarr and Baker, with its several
modifications. Sag methods include the Bloom gelometer, the Cox and
Higby sag method with its variations, the B. A. R. jelly tester, and
Säverborn's Cylindrical torsion method. Christensen was able to show
that the proportion between "sag grade value" and "breaking strength
value" of pectin is not constant but depends upon the molecular weight
of the pectin. These conclusions indicate that it is impossible to derive
a simple conversion factor for the two types of tests.

Sag measurements would appear to have no simple relationship to
sensory evaluations of texture. They are more nearly justifiable on the
basis that they may be able to predict the consumer's visual evaluation
of the sample. Pintauro and Lang (1959) discussed this aspect of the
sag tests in some detail. According to these investigators, measurement
of the change in height of an unmolded jelly gives an incomplete char-

acterization of the gel. The change in the diameter of the base of the gel, the tilting, and other changes in the shape of the gel are important but absent in such simple measurements. Pintauro and Lang were particularly concerned with gelatin desserts served in unmolded form, but certain of their conclusions are applicable to other food gels made with melting type gels such as gelatin. The conclusions are less applicable to non-melting type gels such as agar, alginate, and pectin gels. The solution of Pintauro and Lang to the problem of evaluating thermal sag in gelatin desserts was to obtain a series of shadowgraphs of unmolded gels held at different temperatures.

It would seem to be possible to correlate the results of breaking strength measurements with sensory evaluations of gel texture. Many

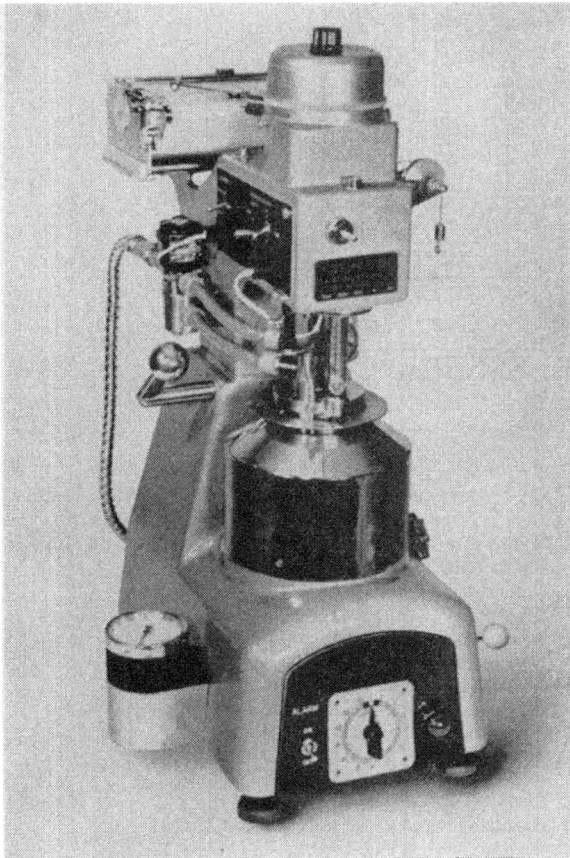

FIG. 14. THE BRABENDER AMYLOGRAPH

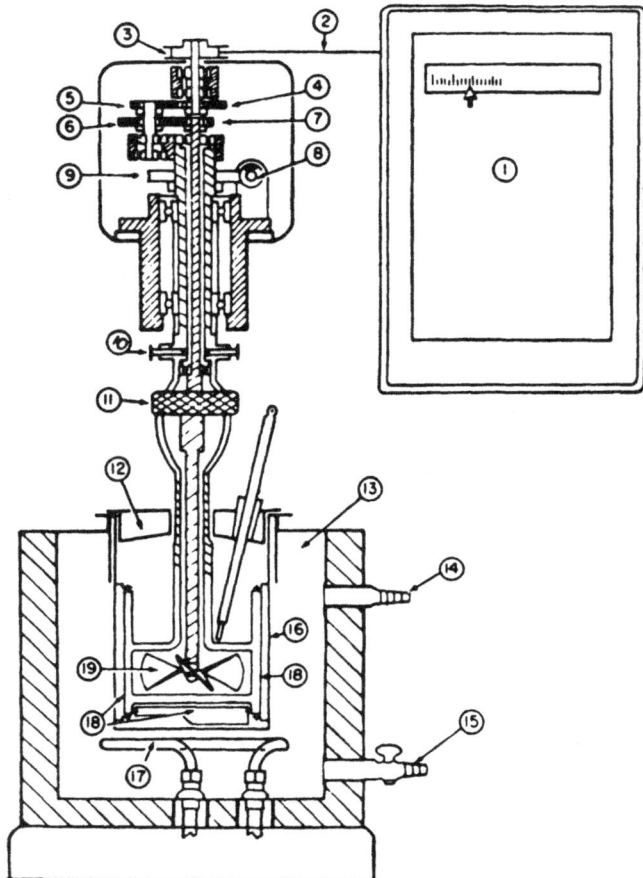

Courtesy Corn Industries Research Foundation

FIG. 15. SECTIONAL DRAWING OF THE CORN INDUSTRIES RE-
SEARCH FOUNDATION VISCOMETER

1. Recorder and dynamometer (dynamometer not shown).
2. Cable from viscometer to recorder.
3. Cable drum.
4, 5, 6, 7. Gears of sun-and-planet differential.
8. Worm, turned by synchronous motor (not shown).
9. Worm gear.
10. Spring pins for holding center shaft.
11. Coupling for attaching stirrer.
12. Condenser and cover.
13. Water bath.
14. Overflow.
15. Drain cock.
16. Starch beaker.
17. Electric heater, thermostatically controlled.
18. Scraper blades.
19. Propeller.

varieties of tests of this general type have been proposed. They frequently involve measurement of the resistance offered by the gel to penetration by a plunger, probe, or paddle. Since the rheological behavior of many gels is time dependent, it might appear on superficial consideration as though it would be preferable to take a sequence of measurements on the samples. So far as food is concerned, however, it is doubtful that the stresses applied in the mouth are of sufficient duration to be much affected by the time dependent potential of the system. Static measurements may actually give a better approximation of the perceived texture.

A more detailed review of methods for determining jelly properties can be found in Joseph and Baier (1949).

Many types of rotational viscometers have been used for measuring consistency of starch gels and sols. Typical of these is the Brabender amylograph illustrated in Fig. 14. The jacket of the bowl of this instrument can be heated or cooled at a predetermined rate of temperature change, thereby facilitating the observation of gelatinization phenomena. This device is much more valuable for the estimation of food texture when it is equipped with a variable speed drive so that consistency measurements can be made at varying shear rates.

The Corn Industries Research Foundation viscometer is another instrument frequently used to study the gelatinization behavior of starches. It is shown diagrammatically in Fig. 15.

BIBLIOGRAPHY

BAKER, G. L. 1938. Improved Delaware jelly strength tester. Fruit Products J. *17*, 329–330.

BAKER, G. L., and GOODWIN, M. W. 1944. Fruit jellies. XII. Effect of methylester content of pectinates upon gel characteristics at different concentrations of sugar. Delaware Univ. Agr. Expt. Sta. Bull. *246*.

BALDWIN, R. R. 1955. Cooking quality correlates with new color test. Rice J. *58*, No. 8, 34.

BARON, M., and SCOTT BLAIR, G. W. 1953. Rheology of cheese and curd. *In* Foodstuffs: Their Plasticity, Fluidity, and Consistency. Edited by G. W. Scott Blair. Interscience Publishers, New York.

BATCHER, O. M., DEARY, P. A., and DAWSON, E. H. 1957. Cooking quality of 26 varieties of milled white rice. Cereal Chem. *34*, 277–285.

BATCHER, O. M., HELMINTOLLER, K. F., and DAWSON, E. H. 1956. Development and application of methods for evaluating cooking and eating quality of rice. Rice J. *59*, No. 13, 4–8, 32.

BATCHER, O. M., LITTLER, R. R., DAWSON, E. H., and HOGAN, J. T. 1958. Cooking quality of white rice milled from rough rice dried at different temperatures. Cereal Chem. *35*, 428–434.

BEACHELL, H. M. 1959. Rice. *In* The Chemistry and Technology of Cereals as Food and Feed. Edited by S. A. Matz. Avi Publishing Co., Westport, Conn.

BEAN, M. L., and OSMAN, E. M. 1959. Behavior of starch during food preparation. II. Effects of different sugars on the viscosity and gel strength of starch pastes. Food Research *24*, 665–671.

BECHTEL, W. G. 1950. Measurements of properties of corn starch gels. J. Colloid Sci. *5*, 260–270.

BINNINGTON, D. S., JOHANNSON, H., and GEDDES, W. F. 1939. Quantitative methods for evaluating the quality of macaroni products. Cereal Chem. *16*, 149–167.

BLISS, C. I., GREENWOOD, M. L., and McKENRICK, M. H. 1953. A comparison of scoring methods for taste tests with mealiness of potatoes. Food Technol. *7*, 491–495.

BLOOM, O. T. 1925. Penetrometer for testing the jelly strength of glues, gelatins, etc. U. S. Pat. 1,540,979. June 9.

BOGGS, M. M., WARD, A. C., SINNOTT, C. N., and KESTER, E. B. 1952. Frozen cooked rice. III. Brown rice. Food Technol. *6*, 53–54.

BOLAFFI, A., MEZZINO, J. F., LOWRY, J. R., and BALDWIN, R. R. 1959. Effects of ionizing radiation on gelatin and the role of various radioprotective agents. Food Technol. *13*, 624–628.

BORASIO, L. 1935. Characteristics of the texture of alimentary pastes. Giorn. risicoltura *25*, 251–257.

BOURNE, E. J., TIFFIN, A. I., and WEIGEL, H. 1960. Interaction of starch with sucrose stearates and other antistaling agents. J. Sci. Food Agr. *11*, 101–109.

BRIMHALL, B., and HIXON, R. M. 1939. The rigidity of starch pastes. Ind. Eng. Chem., Anal. Ed. *11*, 358–361.

BRONSON, W. 1951. Technology and utilization of gelatin. Food Technol. *5*, 51–54.

BROOKS, J. 1960. Strength in the egg. Soc. Chem. Ind. Monograph *7*, 149–177.

BUCHANAN, B. F., and LLOYD, R. L. 1946. Gelatinization of starch. U. S. Pat. 2,406,585. Aug. 27.

CAFFYN, J. E., and BARON, M. 1947. Scientific control in cheese making. Dairyman *64*, 345, 347, 349.

CALLOW, F. H. 1952. Frozen meat. J. Sci. Food Agr. *3*, 146–149.

CHEFTEL, H., and MOCQUARD, J. 1947. The rheological properties of pectin jellies. J. Soc. Chem. Ind. (London) *66*, 297–299.

CHICHESTER, C. O., and STERLING, C. 1957. Measuring stress relaxation in starch gels. Cereal Chem. *34*, 233–237.

CHRISTENSEN, P. E. 1954. Methods of grading pectin in relation to the molecular weight (intrinsic viscosity) of pectin. Food Research *19*, 163–172.

CLUSKEY, J. E., TAYLOR, N. W., and SENTI, F. R. 1959. Relation of the rigidity of flour, starch, and gluten gels to bread staling. Cereal Chem. *36*, 236–246.

COLE, G. M., COX, R. E., and JOSEPH, G. H. 1930. Does sugar inversion affect jelly formation? Food Inds. *2*, 219–221.

COX, R. E., and HIGBY, R. H. 1944. A better way to determine the jellying power of pectins. Food Inds. *16*, 441–442, 505–507.

CULLEN, J. C. 1960. Texture in cooked potatoes. Soc. Chem. Ind. Monograph 7, 128–134.

DAVIS, J. G., HANSON, H. L., and LINEWEAVER, H. 1952. Characterization of the effect of freezing on cooked egg white. Food Research *17*, 393–401.

DAWSON, E. H., MILLER, C., and REDSTROM, R. A. 1956. Cooking quality and flavor of eggs as related to candled quality, storage conditions, and other factors. U. S. Dept. Agr. Agr. Inf. Bull. *164*.

DESIKACHAR, H. S. R., and SUBRAHMANYAN, V. 1959. Expansion of new and old rice during cooking. Cereal Chem. *36*, 385–391.

DESIKACHAR, H. S. R., and SUBRAHMANYAN, V. 1961. The formation of cracks in rice during wetting and its effect on the cooking characteristics of the cereal. Cereal Chem. *38*, 356–364.

DOESBURG, J. J., and GREVERS, G. 1960. Setting time and setting temperature of pectin jellies. Food Research *25*, 634–645.

EMMONS, D. B., PRICE, W. V., and TORRIE, J. H. 1960. Effects of lactic cultures on acidity and firmness of cottage cheese coagulum. J. Dairy Sci. *43*, 480–490.

FERREL, R. E., and KESTER, E. B. 1959. Reduction of cohesion in canned pearl rice by use of edible oil emulsions and surfactants. Food Technol. 13, 473–474.

FETZER, W. R., and KIRST, L. C. 1959. The estimation of starch paste fluidities. Cereal Chem. 36, 108–127.

FORSYTHE, R. H. 1960. Eggs. In Bakery Technology and Engineering. Edited by S. A. Matz. Avi Publishing Co., Westport, Conn.

GHOSH, B. P., and SARKAR, N. 1959. Effect of some inorganic salts on water absorption by rice during cooking. Ann. Biochem. Exptl. Med. (India) 19, No. 3, 83–86.

GOLDSTEIN, A. M. 1954. Chemistry, properties, and application of gum karaya. Advances In Chem. Ser. No. 11, 33–37.

GÖRLING, P., and BEUSCHEL, H. 1959. Shrinkage stresses in drying of gel-like and pasty materials. Chem. Ing. Tech. 31, 393–398.

HALICK, J. V., BEACHELL, H. M., STANSEL, J. W., and KRAMER, H. H. 1960. A note on the determination of gelatinization temperatures of rice varieties. Cereal Chem. 37, 670–672.

HALL, R. C., and FRYER, H. C. 1953. Consistency evaluation of dehydrated potato granules and directions for microscopic rupture count procedure. Food Technol. 7, 373–377.

HAMPEL, G. 1957. The development of staling and the degree of swelling of starch. Getreide u. Mehl 7, 17–22.

HAMPEL, G. 1959. Valuation of rice. Getreide u. Mehl 9, 105–109.

HANNING, F., BLOCH, J. DE G., and SIEMERS, L. L. 1955. The quality of starch puddings containing whey and/or nonfat milk solids. J. Home Econ. 47, 107–109.

HANSON, H. L., CAMPBELL, A. A., and LINEWEAVER, H. 1951. Preparation of stable frozen sauces and gravies. Food Technol. 5, 432–440.

HANSON, H. L., NISHITA, K. D., and LINEWEAVER, H. 1953. Preparation of stable frozen puddings. Food Technol. 7, 462–465.

HARRAP, F. E. G. 1960. Some aspects of the potash nutrition of the potato. J. Sci. Food Agr. 11, 293–298.

HARRINGTON, W. O., OLSON, R. L., and McCREADY, R. M. 1951. Quick-cooking dehydrated potatoes. Food Technol. 5, 311–313.

HARRIS, R. H., and KNOWLES, D. 1943. Macaroni cooking value of some North Dakota durum wheat samples. Food Research 8, 292–298.

HARRIS, R. H., and SIBBITT, L. D. 1958. The cooking properties of some new durum wheat varieties. Food Technol. 12, 91–93.

HARVEY, H. G. 1953. The rheology of certain miscellaneous food products. In Foodstuffs: Their Plasticity, Fluidity, and Consistency. Edited by G. W. Scott Blair. Interscience Publishers, New York.

HARVEY, H. G. 1960. Gels, with special reference to pectin gels. Soc. Chem. Ind. Monograph 7, 29–63.

HEINZE, P. H., KIRKPATRICK, M. E., and DOCHTERMAN, E. F. 1955. Cooking quality and compositional factors of potatoes of different varieties from several commercial locations. U. S. Dept. Agr. Tech. Bull. 1106.

HEISLER, E. G., HUNTER, A. S., WOODWARD, C. F., SICILIANO, J., and TREADWAY, R. H. 1953. Laboratory preparation of potato granules by solvent extraction. Food Technol. 7, 299–302.

HINTON, C. L. 1950. The setting temperature of pectin jellies. J. Sci. Food Agr. 1, 300–307.

HIXON, R. M., and BRIMHALL, B. 1941. A gelometer for starch pastes. Ind. Eng. Chem., Anal. Ed., 13, 193–194.

HOFSTEE, J., and DE WILLIGEN, A. H. A. 1950. Proportion of gels formed in pastes at various stages of heating. Chem. Weekblad 46, 649.

HOFSTEE, J., and DE WILLIGEN, A. H. A. 1953. Starch. In Foodstuffs: Their Plasticity, Fluidity, and Consistency. Edited by G. W. Scott Blair. Interscience Publishers, New York.

HOGAN, J. T., and PLANCK, R. W. 1958. Hydration characteristics of rice as influenced by variety and drying method. Cereal Chem. *35*, 469–482.

HOOGZAND, C., and DOESBURG, J. J. 1961. Effect of blanching on texture and pectin of canned cauliflower. Food Technol. *15*, 160–163.

HUET, R. 1956. Note on the biochemical significance of firmness in the case of the pulp of the banana. Fruits *11*, 395–399.

IDSON, B., and BRASWELL, E. 1957. Gelatin. Advances in Food Research 7, 235–338.

JACOBS, M. B. 1959. The Chemistry and Technology of Food and Food Products. Third Edition. Interscience Publishers, New York.

JONGH, G. 1961. The formation of dough and bread structures. I. The ability of starch to form structures and the improving effect of glyceryl monostearate. Cereal Chem. *38*, 140–152.

JOSEPH, G. H., and BAIER, W. E. 1949. Methods of determining the firmness and setting time of pectin test jellies. Food Technol. *3*, 18–22.

JOUX, J. L. 1957. Role of pectic substances in the maintenance of firmness in pasteurized apricots. Compt. rend. acad. agr. France *43*, 506–513.

KERR, R. W. 1950. Chemistry and Industry of Starch. Academic Press, New York.

KERTESZ, Z. I., MORGAN, B. H., TUTTLE, L. W., and LAVIN, M. 1956. Effect of ionizing radiations on pectin. Radiation Research *5*, 372–381.

KESLER, C. C., and BECHTEL, W. G. 1947. Recording viscometer for starches. Anal. Chem. *19*, 16–19.

KUHN, G., DESROSIER, N. W., and AMMERMAN, G. 1959. Relation of chemical composition and some physical properties to potato texture. Food Technol. *13*, 183–185.

KULP, K., and BECHTEL, W. G. 1961. Effect of freezing and frozen storage on the freshness and firmness of Danish pastry. Food Technol. *15*, 273–275.

LABELLE, R. L., SHALLENBERGER, R. S., WAY, R. D., MATTICK, L. R., and MOYER, J. C. 1960. The relationship of apple maturity to applesauce quality. Food Technol. *14*, 463–468.

LANCASTER, E. B., and ANDERSON, R. A. 1959. Consistency measurements on batters, doughs, and pastes. Cereal Chem. *36*, 420–430.

LEVINE, A. S., FELLERS, C. R., and BARTON, R. R. 1950. Preservation of Russian caviar by canning. Food Technol. *4*, 15–16.

LITTLE, R. R., and HILDER, G. B. 1960. Differential response of rice starch granules to heating in water at 62°C. Cereal Chem. *37*, 456–463.

LONGRÉE, K. 1950. Quality problems in cooked, frozen potatoes. Food Technol. *4*, 98–104.

MACMASTERS, M. M., and WOLFF, I. A. 1959. Characteristics of cereal starches. *In* The Chemistry and Technology of Cereals as Food and Feed. Avi Publishing Co., Westport, Conn.

MANTELL, C. L. 1954. Technology of gum arabic. Advances in Chem. Ser. No. *11*, 20–32.

MATELES, R. I., and GOLDBLITH, S. A. 1958. Some effects of ionizing radiations on gelatin. Food Technol. *12*, 633–640.

MATTSON, S. 1946. The cookability of yellow peas. A colloichemical and biochemical study. Acta Agr. Suecana *2*, 185–231.

MATTSON, S., ÅKERBERG, E., ERICKSSON, E., KOUTLER-ANDERSSON, E., and VAHTRAS, K. 1950. Factors determining the composition and cookability of peas. Acta Agr. Scand. *1*, 40–61.

MEYER, K. H., BERNFELD, P. BOISSONNAS, R. A. GÜRTLER, P., and NOELTING, G. 1949. Starch solutions and pastes and their molecular interpretation. J. Phys. Colloid Chem. *53*, 319–325.

MIYAKE, M., and HAYASHI, K. 1956. Studies on fish jellies. II. Content of the myosin fraction in fish muscle. Bull. Japan. Soc. Sci. Fisheries *22*, 48–51.

MORSE, L. M., DAVIS, D. S., and JACK, E. L. 1950. Use and properties of nonfat dry milk solids in food preparation. I. Effect on viscosity and gel strength. Food Research 15, 200–215.

MOYLS, A. W., ATKINSON, F. E., STRACHAN, C. C., and BRITTON, D. D. 1955. Preparation and storage of canned berry and berry-apple pie fillings. Food Technol. 9, 629–634.

NAIK-KURADE, A. G., LIVINSTON, G. E., FRANCIS, F. J., and FAGERSON, I. S. 1959. Effects of cathode ray and gamma ray irradiation on some organic acid-carbohydrate systems. Food Research 24, 618–632.

OKADA, M., and YAMAZAKI, A. 1956. Relationship between jelly strength and chemical composition of fish meat jelly. Bull. Tokai reg. Fish. Res. Lab. Tokyo 13.

OKAMURA, K., MATUDA, T., and YOKOYAMA, M. 1958. Studies on the action of phosphates in "kamaboko" and fish sausage products. I. Various effects of phosphate on qualities of "kamaboko" fish cake. Bull. Japan. Soc. Sci. Fisheries 24, 545–550.

OLLIVER, M. 1950. Factors affecting the jelly grading of pectins. Food Technol. 14, 370–375.

OLLIVER, M., WADE, P., and DENT, K. 1957A. The determination of grade strength of pectins by the Teepol-gel procedure. Analyst 82, 127–128.

OLLIVER, M., WADE, P., and DENT, K. 1957B. The jelly strength grading of pectins for use in jam manufacture. J. Sci. Food Agr. 8, 188–196.

OLSEN, A. G. 1934. Pectin studies. III. General theory of pectin jelly formation. J. Phys. Chem. 38, 919–930.

OLSEN, A. G., STUEWER, R. F., FEHLBERG, E. R., and BEACH, N. M. 1939. Pectin studies. Relation of combining weight to other properties of commercial pectins. Ind. Eng. Chem. 31, 1015–1020.

OLSON, N. F. 1959. Study of the control of the physical structure of pasteurized process cheese spreads. Dissertation Abstr. 19, 2701–2702.

OWENS, H. S., SWENSON, H. A., and SCHULTZ, T. S. 1954. Factors influencing gelation with pectin. Advances in Chem. Ser. No. 11, 10–15.

PAULSEN, T. M. 1961. A study of macaroni products containing soy flour. Food Technol. 15, 118–121.

PERSONIUS, C. J., and SHARP, P. F. 1938. Adhesion of potato tissue cells as influenced by pectic solvents and precipitants. Food Research 4, 299–305.

PINTAURO, N. D., and LANG, R. E. 1959. Graphic measurements of unmolded gels. Food Research 24, 310–318.

POWERS, J. J., PRATT, D. E., and JOINER, J. B. 1961. Gelation of canned peas and pinto beans as influenced by processing conditions, starch, and pectic content. Food Technol. 15, 41–47.

PYKE, W. E., and JOHNSON, G. 1940. The relation of the calcium ion to the sloughing of potatoes. Am. Potato J. 17, 1–9.

RANKIN, J. C., MEHLTRETTER, C. L., and SENTI, F. R. 1959. Hydroxyethylated cereal flours. Cereal Chem. 36, 215–227.

RANKIN, J. C., RALL, J. G., RUSSELL, C. R., and SENTI, F. R. 1960. Preparation and properties of hydroxyethylated high-amylose corn starch. Cereal Chem. 37, 656–670.

RAO, B. S. 1948. Rice. Physico-chemical aspects of its curing to secure vitamin conservation and improvement in storage and cooking quality. Proc. Indian Sci. Congr. (Patna) 35, 1–12.

RAO, B. S., VASUVEDA MURTHY, A. R. ,and SUBRAMANYA, R. S. 1952. The amylose and amylopectin contents of rice and their influence on the cooking quality of the cereal. Proc. Indian Acad. Sci. 36B, 70–80.

REEVE, R. M., and NOTTER, G. K. 1959. An improved method for counting ruptured cells in dehydrated potato products. Food Technol. 13, 574–577.

RHODES, W. E., and DAVIES, H. F. 1945. The selection and pre-processing of potatoes for canning with special reference to control of texture by calcium chloride. Chem. Ind. 21, 162–163.

84 FOOD TEXTURE

84   FOOD TEXTURE

84   FOOD TEXTURE

ROSEMAN, A. S. 1958. The effect of freezing on the hydration characteristics of rice. Food Technol. 12, 464–468.

SALWIN, H., BLOCH, I., and MITCHELL, J. H., JR. 1953. Dehydrated stabilized egg, importance and determination of pH. Food Technol. 7, 447–452.

SCHARSCHMIDT, R. K. 1954. Factors affecting the sloughing of canned potatoes. Masters Thesis, University of Wisconsin.

SCHOCH, T. J., and ELDER, A. L. 1955. Starches in the food industry. Advances in Chem. Ser. No. 12, 21–34.

SCHWARZ, T. W., and LEVY, G. 1957. Viscosity changes of sodium alginate solutions after freezing and thawing. J. Am. Pharm. Assoc., Sci. Ed., 46, 562–563.

SCOTT BLAIR, G. W. 1960. Rheology as an aid to the specification of cheese types. Dairy Inds. 25, 585–587.

SELBY, H. H. 1954. Agar since 1943. Advances in Chem. Ser. No. 11, 16–19.

SHALLENBERGER, R. S., and MOYER, J. C. 1961. Sugar-starch transformations in peas. J. Agr. Food Chem. 9, 137–140.

SHIMAZU, F., and STERLING, C. 1961. Dehydration in model systems: Cellulose and calcium pectinate. J. Food Sci. 26, 291–296.

SHIMIZU, T., FUKAWA, H., and ICHIBA, A. 1958. Physical properties of noodles. Cereal Chem. 35, 34–46.

SIMON, M., WAGNER, J. R., SILVEIRA, V. G., and HENDEL, C. E. 1953. Influence of piece size on production and quality of dehydrated Irish potatoes. Food Technol. 7, 423–428.

SMITH, A. K., WATANABE, T., and NASH, A. M. 1960. Tofu from Japanese and United States soybeans. Food Technol. 14, 332–336.

SREENIVASAN, A. 1939. Studies in quality on rice. IV. Storage changes in rice after harvest. Indian J. Agr. Sci. 9, 208–222.

STEINER, A. B., and MCNEELY, W. H. 1954. Algin in review. Advances in Chem. Ser. No. 11, 68–82.

STERLING, C. 1957A. Structure of oriented gels of calcium polyuronates. Biochim. Biophys. Acta 26, 186–197.

STERLING, C. 1957B. Relaxation of stress in starch jelly candy. Cereal Chem. 34, 238–246.

STERLING, C. 1957C. Retrogradation in a starch jelly candy. Food Research 22, 184–191.

STERLING, C. 1958. The effect of temperature on the "setting" of starch gels. Mfg. Confectioner 28, No. 2, 21–22, 30.

STIER, E. F., BALL, C. O., and MACLINN, W. A. 1956. Changes in pectic substances of tomatoes during storage. Food Technol. 10, 40–43.

STOLOFF, L. 1954. Irish moss extractives. Advances in Chem. Ser. No. 11, 92–100.

UNRAU, A. M., and NYLUND, R. E. 1957. The relation of physical properties and chemical composition to mealiness in the potato. I. Physical properties. Am. Potato J. 34, 245–253.

VEIS, A., and COHEN, J. 1960. Reversible transformation of gelatin to the collagen structure. Nature 186, 720–721.

VON HIPPEL, P. H., GALLOP, P. M., SEIFTER, S., and CUNNINGHAM, R. S. 1960. An enzymatic examination of the structure of the collagen molecule. J. Am. Chem. Soc. 82, 2774–2786.

WARD, A. G., and SAUNDERS, P. R. 1956. Rheology of gelation. In Rheology; Theory and Applications. Academic Press, New York.

WEBB, B. H., and HUFNAGEL, C. F. 1946. The effect of milk products on the heat-stability and viscosity of cream-style foods. J. Dairy Sci. 29, 221–230.

WECKEL, K. G., SCHARSCHMIDT, R. K., and RIEMAN, G. H. 1959. Sloughing in canned potatoes. Food Technol. 13, 456–459.

WEST, L. C., TITUS, M. C., and VAN DUYNE, F. O. 1959. Effect of freezer storage and variations in preparation on bacterial count, palatability, and thiamine content of ham loaf, Italian rice, and chicken. Food Technol. 13, 323–327.

WHISTLER, R. L.   1954.   Guar gum, locust bean gum, and others.   Advances in Chem. Ser. No. *11*, 45–50.

WHISTLER, R. L., and BEMILLER, J. N.   1960.   Alkaline degradation of alginates. J. Am. Chem. Soc. *82*, 457–459.

WHITTENBERGER, R. T.   1951.   Changes in specific gravity, starch content, and sloughing of potatoes during storage.   Am. Potato J. *28*, 738–747.

WOOD, G. C.   1960.   The formation of fibrils from collagen solutions.   II.   A mechanism of collagen-fibril formation.   Biochem. J. *1960*, 598–605.

WOODMANSEE, C. W., McCLENDON, J. H., and SOMERS, G. F.   1959.   Chemical changes associated with the ripening of apples and tomatoes.   Food Research *24*, 503–514.

CHAPTER 5

Fibers

INTRODUCTION

Very many natural and manufactured foods contain fibrous structures which are the most important contributors to their texture. Muscle tissues and vegetables such as celery and asparagus are examples which come to mind immediately. On closer examination, it can be seen that virtually all fruits and vegetables contain some components which are perceived as fibers when the food is eaten.

In this chapter, the nature of the fibers themselves and the way in which they affect texture perception will be examined, the literature on measurement methods will be critically reviewed, and the effect of different conditions on the texture quality of fibrous foods will be discussed.

VEGETABLE TISSUES

The fibers which form one of the most important texture-influencing constituents in many vegetables are composed of cells rich in cellulose. Frequently they follow a development pattern beginning with relatively small, easily separated cells having thin walls and pectinous middle lamellae in immature examples to well-consolidated groups of thick walled cells forming long coarse and woody strings in the most mature specimens.

Isherwood (1955 and 1960) called attention to the role of lignin in toughening fibrous elements and preventing them from softening and disintegrating during cooking. Less than two per cent of lignin markedly influences swelling and solubility of the polysaccharides of the cell wall. Investigation of the role of lignin in runner beans indicated that lignification of cellulose fibers in the parchment layer was responsible for stringiness in older beans. The extent of lignification varies with the season and may be increased by any drying of the pod.

Klimczak and Piekarska (1957) reported that the content of cellulose increased in kohlrabi during growth but did not vary much in carrots, spinach, and radishes gathered at different stages of development.

According to Kramer and Backinger (1959), the toughness of pods in green and wax beans is caused by the development of oblique fibers in the pod walls. Some varieties develop these to a greater extent than others. Fiber content is also affected by growing conditions, especially by temperature, precipitation, and stage of maturity. Fibrousness is a particularly serious problem with flat beans, but may also be encountered with round varieties harvested at late maturity, especially under hot dry conditions.

Another study directly concerned with the relationship of fiber development to quality of snap bean pods was reported by Stark and Mahoney (1942). They stated that the fibrous sheath in the side wall of the pods consisted mainly of mesocarp. Increase in width of the inner mesocarp occurs at a constant rate until the large sieve sizes of beans are reached. External factors such as climatic conditions appear responsible for the amount of thickening in the cell wall of the varieties studied.

Caldwell and Culpepper (1943A and 1943B) and Caldwell *et al.* (1944) studied texture of snap beans before and after dehydration and reconstitution. Texture was found to be a function mainly of the degree of toughness or tenderness of the pericarp. Extremes of toughness were associated with "woodiness" or fibrousness. These characteristics were due to inherent qualities of the variety and were not much changed by any of the treatments studied.

Although they are sometimes used on products such as asparagus and snap beans, pressure testers incorporating a narrow pointed or rounded plunger probably cannot give an accurate measurement of fibrousness. Culpepper and Moon (1935) used such an instrument (described earlier by Culpepper and Magoon 1924) for testing the toughness of asparagus. This device was a pressure tester equipped with a very thin plunger. It seems obvious that a thin probe would tend to spread the fibers apart instead of cutting them. Guyer and Kramer (1951) got around this difficulty by replacing the plunger on a pressure tester for fruit with a piece of straight stainless steel having a blunt cutting edge 0.030 of an inch thick.

Pressure testers fitted with needle-like probes could, of course, be used to gain valuable information about the non-fibrous components of texture. MacGillivray (1933) used a Magness and Taylor tester with a modified plunger one-eighth inch in diameter to evaluate asparagus tenderness, after he had scraped off the outside portion of the stalk with a knife, thereby removing the lignified pericycle fibers and exposing the inner whitish tissue.

Most investigators have inclined to the use of shear-type actions for measuring fibrousness. Wilder (1948) described a "fibrometer" and a method for its use which made it possible to record the number of tough spears in a sample of asparagus, and, for each individual spear, the length of the fibrous portion. In this method, each spear is tested at regular intervals along its length by a standardized cutting device. The essential part of the cutter is a wire held rigidly in a frame so arranged that the horizontal wire may be lowered upon the supporting asparagus unit and allowed to press down with a definite shearing force. Presence of fiber is indicated by a failure to cut through the spear. The principle on which this device is based seems sound enough and it was later adapted and modified by other investigators for measuring texture in other fibrous vegetables.

Gould (1949) adapted Wilder's asparagus fibrometer to the quality testing of snap and wax beans. He found that it was necessary to be able to apply variable forces to the cutting edge. Use of different weights applied to the lever in succession was unsatisfactory, and he finally settled upon a gage for measuring resistance offered by the beans to the cutting force. The instrument in its improved form used the sample holder box of the fibrometer, but had seven wires 0.031 of an inch in diameter as the cutting means. An air-cushion device was employed to regulate the lowering speed. Using this apparatus, Gould (1950A, 1950B, and 1950C) studied the maturation of snap beans. Tenderness and fiber were found to be highly correlated with maturity. Texturometer values of the beans had a significant correlation coefficient of -0.55 with maturity. Per cent by weight of fiber for U. S. Department of Agriculture grades were 0.05 per cent for A, 0.10 per cent for B, and 0.15 per cent for C.

Kramer and his associates have published extensively on the development of shear-presses for determining fibrousness and other textural qualities of foods. An early publication (Kramer *et al.* 1951) described an instrument consisting of a hydraulic pump, gage, cylinder, piston, and test cell. A different test cell was used for each product. For beans, a series of square-edged bars less than one-eighth of an inch thick was pressed through a sample held between two grids. Several pods were measured at one time, tending to average out the within-batch variability. Some variables between different machines which were found to affect results were: (1) speed of piston travel, and (2) design of the test cell, especially the tolerance between the grid bars and the slits at the bottom of the test cells. When lima beans were tested with the device, values had a significant correlation coefficient of 0.87 with U. S. grades.

Improvements made to the Kramer press (Kramer and Aamlid 1953) shortly after its introduction included an increase in the capacity of the hydraulic pump, an improved valving arrangement, the addition of a metering device to control piston speed, and the replacement of the hydraulic gage by a more precise strain gage. The importance of developing a specialized test cell for each type of product was re-emphasized. Two different cells were used in a study on peas. One of these was very similar to a cell previously used for lima beans while the other had a solid, instead of slotted, bottom. In the latter case, the moving bars were driven no closer than three-sixteenths of an inch from the bottom. The shear-press gave results equal in precision and accuracy to the tenderometer and superior to the texturometer, the Cefaly instrument, and the U. S. Department of Agriculture raw grade.

Still further improvements were described by Kramer and Backinger (1959) and Orchard (1961). The instrument now includes a "proving ring dynamometer" into which is fitted the measuring element. One of the available test cells is similar to the fiber pressure tester. It simulates the cutting action of the teeth and is used to measure the fibrousness of products such as asparagus and okra. Kramer and Backinger found a highly significant correlation coefficient of 0.93 between shear-press determinations on raw flat or round green beans and the fiber content of the processed samples. A reading of 1,270 lbs. was found to be equivalent to the maximum fiber allowed by the U. S. Department of Agriculture.

Backinger et al. (1957) used time-force curves based on data from shear-press measurements as indications of fibrousness in asparagus. These curves provided additional and more precise information on fibrousness than could be deduced from the maximum force measurements alone. They found that average fibrousness could be indicated in terms of work done per square inch of sample, at a given speed of stroke. Stalk diameter and sample stratification had some effect on maximum force values, but the effects on work values were not usually as great. Variability within the lot was estimated from an examination of the work curves.

A modified shear press which could be used for measuring the fibrousness of asparagus and string beans was discussed by Kramer (1957). This device, with still more improvements, was the subject of a symposium held in 1961 (Orchard 1961).

Van Buren et al. (1960) studied the firmness of canned snap beans using a different type of texturometer by which a measurement is made of the force required to move a piston seven cm. in diameter to a distance of 3.8 cm. from the bottom of a can of drained beans. Strain gages and a recording potentiometer were used to indicate the force.

Quality standards of the Food and Drug Administration provide that garden and wax beans must contain less than 0.15 per cent fiber as measured by a chemical procedure involving the digestion of a sample of beans in boiling water and lye, and disintegration in a malted milk stirrer. Any material retained by a 30-mesh screen is considered to be fiber (Rowe and Bonney 1936).

Kramer and Smith (1945) discussed a blendor disintegration method for estimating fiber in green beans, and somewhat later Guyer and Kramer (1951) compared the results obtained when fiber in green and wax beans was measured by a digestion method and by a method comprising disintegration in a blender, washing the residue on a 30-mesh screen, drying, and weighing. Pressure fiber recordings were also made across the pods at a right angle to the suture using a modified fruit tester which had had the plunger replaced by a piece of stainless steel with a blunt cutting edge 0.030 of an inch thick. The results of the objective tests were compared with expert panel evaluations. They concluded that fiber content increased as the beans matured. There was also a pronounced varietal difference in fiber content.

When snap beans were stored up to ten days at 35°, 50°, or 70°F., there was no detectable change in the fiber content as measured by the digestion method, but the expert panel and the disintegration method results indicated that fiber had increased. The disintegration method appeared to be at least equal in accuracy to the digestion method, and it was simpler to perform.

MacGillivray (1933) tested stalks of asparagus using a procedure which in effect eliminated the fibrous component of texture, and found that spears harvested during the first third of the season were tougher than those harvested later. Shewfelt and Mohr (1960) found that a higher fiber content in asparagus was associated with the longer spears and with the butt portion of the spear. Subjective texture ratings were better for the shorter spears. However, they found no significant variation in texture due to harvest date.

Although texture in peas is, to a large extent, a function of the cell characteristics and the moisture content of the cotyledons, the fibrousness of the seed coat undoubtedly plays a significant role in perceived texture. The "tenderometer" (Martin et al. 1938) has been much used for grading texture in peas. In one version of this device, the product is sheared through a fixed grid by an assembly of bars (or a grid) slowly rotated by a motor. A weighted pendulum and pointer attached to the fixed grid provide an indication of the torque transmitted through the sample (Martin 1937A and B). It has been found that there is a significant relationship between the tenderometer readings on raw peas and the commercial grade of the product. Sayre (1954) compared the effectiveness of the tenderometer and the maturometer for measuring the quality of raw peas. The latter instrument determines the mass resistance of 143 peas to puncture by steel rods. Readings from the two instruments showed a high positive correlation, and both instruments gave values which were correlated with the per cent of alcohol-insoluble solids in the peas (fiber is included in this fraction).

MEATS

In the literature of texture measurement, meat studies occupy more space than those of any other group of foods. This situation is probably the direct result of the universally recognized economic importance of tenderness in this relatively high-priced food. Several investigators (e.g., Kropf and Graf 1959A; Rhodes *et al.* 1955; and Sperring *et al.* 1956) have concluded that tenderness is probably the single most important factor affecting consumer evaluation of meat quality and acceptability.

Muscles are composed of a large number of fibers gathered together in bundles lying approximately parallel to one another in a matrix of connective tissue. Fundamentally, the fibers are composed of elastic proteins and possess many gel-like characteristics. Fatty inclusions are found in muscles in quantities and dimensions dependent upon the condition of the animal and the location of the muscle.

It is clear that fibrousness is a very important textural factor in meats but that it is not the only contributor to perceived texture in these products. The toughness of the fascia, the proportion of fat, and the quality and amount of collagen are some of the factors other than fiber characteristics which might be expected to influence tenderness.

On the other hand, Paul *et al.* (1956) in a study conducted with commercial grade beef from cows discovered that there was no relationship between juiciness and tenderness or total moisture content, or between fat content and tenderness. In this type of meat, at least, fiber characteristics must have been the only major factors influencing tenderness. Cover *et al.* (1958) also found a very low level of relationship between fat content and the tenderness of beef, whether the latter was determined by taste panel or by Warner-Bratzler shear.

Sale (1960) divided tenderness-measuring devices into four types depending upon their mode of action. These classes were: (1) simple shear, (2) penetration, (3) biting, and (4) mincing. There is considerable overlapping when this classification is applied to the instruments described in the literature, but it is probably as good as any other available scheme.

Any texture-measuring procedure for meats which purports to correlate well with organoleptic evaluations must include means for determining the resistance to shear. Although pressure testers equipped with probes of small diameter have been used for measuring the texture of muscle tissue, it would appear that such devices would be rather ineffective in giving an indication of the fibrousness of samples. If the rod is sufficiently blunt, as in the apparatus of Tressler *et al.* (1932), worthwhile values might be obtained. The puncturing part of the latter device was five-sixteenths of an inch in diameter.

Bratzler (1932) described a method for measuring the tenderness of meat by means of mechanical shear which was made the basis for the well-known Warner-Bratzler (Bratzler 1949) shear testing procedure. Deatherage and Garnatz (1952) concluded that measurements obtained by the Warner-Bratzler apparatus were not closely related to a taste panel's evaluation of the tenderness of broiled steaks from matched pairs of short loins. Sleeth *et al.* (1957) similarly found that Warner-Bratzler shear values correlated poorly with panel scores for tenderness. Visser *et al.* (1960) state that shear values were not significantly correlated with tenderness scores for cooked beef obtained in their investigation. On the other hand, Saffle and Bratzler (1959) found low but highly significant correlations between taste panel scores for overall acceptability of loin pork chops and shear (as well as the muscle fiber extensibility). Low but highly significant correlations were also observed between muscle fiber extension and both shear and marbling. Kropf and Graf (1959A and 1959B) compared panel evaluations of the texture of beef with grade attributes and the results obtained by several objective tests, including Warner-Bratzler shear. The panels were asked to rate texture and "firmness" separately. Texture was rated on a 7-point scale ranging from "Very fine" to "Extremely coarse" while firmness was rated from "Very firm" to "Extremely soft" on a 7-point scale. The texture of lean showed a highly significant correlation coefficient with color of lean ($r = 0.84$) and bone ossification ($r = 0.64$) and had a significant r of -0.54 to the mechanical shear value. Sensory tenderness (firmness) had a highly significant correlation ($r = -0.78$) with mechanical shear values. Lower shear values, as well as higher tenderness ratings and higher ether extract values, were associated with higher carcass grades. For a summary of the results of Kropf and Graf, see Table 8.

Miyada and Tappel (1956) devised a method for estimating meat tenderness by measuring the energy consumed in grinding a sample of the food in a household meat grinder. Using essentially the same method, Bockian *et al.* (1958) tested samples from the longissimus dorsi muscles of standing rib roasts and compared the values so obtained with subjective evaluations of tenderness by trained judges. A significant coefficient of -0.60 was obtained between the objective and subjective tests. Duplicate samples showed about a ten per cent variation in the energy required to grind them.

Simone *et al.* (1959) modified the Miyada and Tappel grinder method by improving the recording circuit. A one ohm resistance was inserted in series with the grinder leads, and the voltage drop across this was measured. The AC voltage drop was rectified by four crystal diodes in a full-wave bridge whose output was fed to a resistance-capacitance load

TABLE 8

CORRELATION COEFFICIENTS OF VARIOUS SUBJECTIVE, CHEMICAL, AND SENSORY EVALUATIONS OF MEAT QUALITY[1]

Length of Carcass 1	Round Index[2] 2	Fat Covering[3] 3	Marbling[3] 4	Texture of Lean[3] 5	Firmness of Lean[3] 6	Color of Lean[3] 7	Bone Ossification[3] 8
	0.05	0.06	0.34	−0.77[5]	0.12	−0.49[4]	−0.76[5]
		0.89[5]	0.87[5]	0.36	0.90[5]	0.66[5]	−0.19
			0.84[5]	0.20	0.87[5]	0.48[4]	−0.30
				0.03	0.90[5]	0.42	−0.49[4]
					0.24	0.84[5]	0.64[5]
						0.56[4]	−0.43
							0.40

Carcass Weight 9	Sensory Preference 10	Sensory Tenderness 11	Sensory Flavor 12	Sensory Juiciness 13	Mechanical Shear Value 14	Ether Extract of Loin Eye 15	vs.
0.84[5]	−0.55[4]	−0.07	−0.31	−0.03	0.18	0.57[4]	1
0.44	0.07	0.53[4]	0.40	0.35	−0.83[5]	0.48[4]	2
0.35	0.28	0.66[5]	0.48[4]	0.24	−0.79[5]	0.61[5]	3
0.60[5]	−0.03	0.62[5]	0.26	0.37	−0.79[5]	0.74[5]	4
−0.43	0.41	0.29	0.36	0.29	−0.54[4]	−0.28	5
0.37	0.07	0.68[5]	0.23	0.34	−0.85[5]	0.65[5]	6
−0.08	0.21	0.35	0.45	0.25	−0.74[5]	0.06	7
−0.50[4]	0.23	−0.32	0.31	0.02	0.06	−0.64[5]	8
	−0.53[4]	0.06	−0.04	0.20	−0.20	0.59[4]	9
		0.53[4]	0.43	0.17	−0.27	−0.07	10
			0.23	0.53[4]	−0.78[5]	0.54[4]	11
				0.18	−0.39	0.26	12
					−0.54[4]	0.24	13
						−0.47[4]	14

[1] From Kropf and Graf (1959A).
[2] Round index is the ratio of the circumference of the round to its length.
[3] A subjective estimate.
[4] Probability less than five per cent.
[5] Probability less than one per cent.

having a time constant ∼0.25 seconds. The resultant output voltage was equal to that developed when the grinder was run without a load. The difference signal was then fed to a one millivolt Bristol dynamometer recorder. The chart obtained was cut along the curve and the area estimated by weighing the cut-out. Tenderness evaluations by this method were not as precise as panel scoring.

Emerson and Palmer (1960) used a Hamilton-Beach household food grinder wired in series with a strip-chart recording ammeter. The input voltage was held constant, and the total ampere-seconds determined by measuring the area under the ampere-time charts. Joules of energy per gram of sample (E/g) were calculated by applying the formula

$$\frac{\text{Ampere seconds} \times \text{volts}}{\text{Sample weight in grams}} = \frac{E}{g}$$

and used as an indication of the sample tenderness. In addition, to these determinations, tenderness of loin steaks and rib roasts was evaluated by a taste panel and the Warner-Bratzler shear. Two sets of experiments were reported, and the correlations varied considerably between them, but, in general, the taste panel seemed to yield the most reproducible data. These workers concluded,

Under the conditions of this investigation, the Warner-Bratzler shear apparatus appears to be a more precise measurement of tenderness than the food grinder.

A modified Carver juice press has been used to estimate the tenderness of beef muscle (Sperring *et al.* 1956). Meat plugs one-half inch in thick-

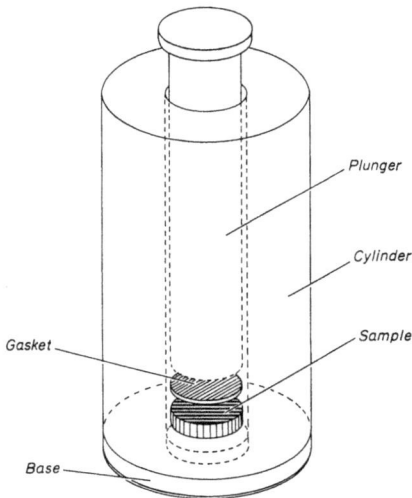

Fig. 16. Diagram of Extrusion Cylinder for Meat Samples

ness were placed in a cylinder 2.8 cm. in inside diameter and having one hole 0.3 cm. in diameter leading to the outside from the bottom. The meat was covered with a rubber gasket and a piston was driven down upon it by constantly increasing pressure. The reading on the hydraulic pressure gage was taken when the first protrusion of meat was observed. The values obtained with this "tenderness press" were significantly correlated with Warner-Bratzler shear data and panel ratings. It is rather difficult to estimate the contribution of fiber quality to this measurement, but it must be of considerable influence.

An apparatus for measuring the tensile strength of individual fibers was developed by Smith (1957). Although he found that there was no great difference in the breaking strength of single fibers from different muscles, he apparently did not attempt to correlate the results obtained with his apparatus and subjective tenderness evaluations.

The apparatus of Smith and the Warner-Bratzler shear method were used to test beef muscle and the results were compared with taste panel evaluation by Bramblett *et al.* (1959). The tensile strength of single fibers was not significantly correlated with the shear or with panel results obtained in this investigation. However, see Fig. 17 summarizing data of May *et al.* (1962).

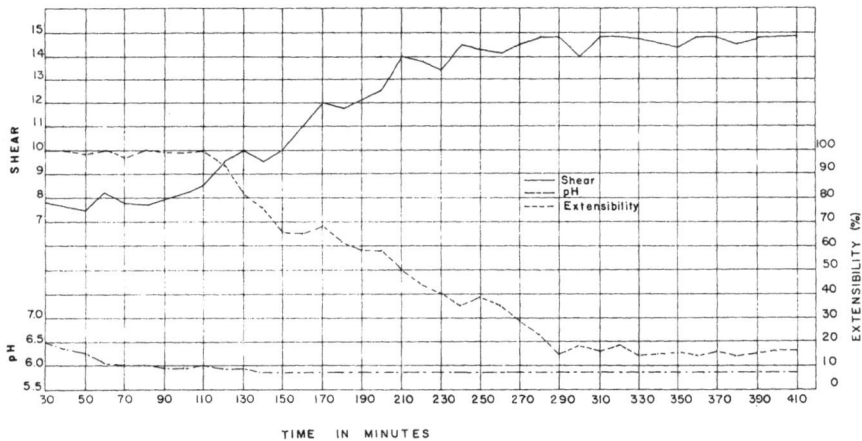

FIG. 17. RELATIONSHIP OF pH, SHEAR, AND EXTENSIBILITY OF A TYPICAL PORK MUSCLE AFTER SLAUGHTER

Chemical methods have not been very successful in predicting meat tenderness. Of several biochemical factors studied by Husaini *et al.* (1950) only alkali-insoluble protein and muscle plasma as represented by muscle hemoglobin appeared to be correlated with changes in the tenderness of beef. Lloyd and Hiner (1959) observed a significant correlation between the total hydroxyproline content of three fractions of alkali-insoluble protein from beef muscle and two measurements of tenderness. Lörincz and Szeredy (1959) extracted muscles with 0.05 N alkali at room temperature for 24 hours, and then compared the nitrogen in the extract with total nitrogen. The difference was described as connective tissue and was said to be related to resistance to chewing.

Love (1960) reported a method for direct estimation of the toughness of fish flesh. A sample was cut so as to obtain a constant proportion of intact cells and a weighed portion was agitated in aqueous formaldehyde for a fixed period (20 to 30 seconds). The optical density of the slurry, which was read in a colorimeter, increased with the period of storage of the frozen fish as a result of the progressive toughening of the cells.

Nearly all investigators who have studied the problem agree that, within the same breed and under the same conditions of feeding and management, muscle tissues from older animals are not as tender as similar muscle tissues from young animals. For example, Peterson *et al.* (1959) observed this effect in chickens. Dunsing (1959 A and B) found that the age of beef animals significantly affected the texture of steaks. Hiner *et al.* (1953) observed a significant increase in fiber diameter with age in all beef muscles studied. Histological examinations (Lörincz and Szeredy 1959) have shown that the connective tissue fibers of older animals are thicker, more twisted, less uniform in size, and more closely woven together. All of these changes contribute to increased toughness.

Conditions of post mortem aging have a critical effect upon the tenderness of most kinds of meat. In chickens and turkeys, the age of the birds at time of slaughter, the duration of post mortem aging, the class of poultry, the temperature of aging, and the media in which the carcasses were aged, all appear to be important factors in post mortem tenderization (Dodge and Stadelman 1959). Relaxed muscles are tenderer than partly contracted muscles (Locker 1960). Tenderization is discussed further in the chapter on the effect of enzymatic reactions on food texture.

There is a difference between the tensile strength of single fibers from the same muscle location in different carcasses and from different muscles from the same carcass (Bramblett *et al.* 1959). For example, beef adductor and gracilis muscles are uniformly more tender than biceps femoris, semimembranosus, or semitendinosus. Furthermore, it has been shown that tenderness varies from one end of the muscle (loin) to the other (Ramsbottom *et al.* 1945 and Simone *et al.* 1959). In chicken breast muscle (pectoralis major) resistance to shear has been shown (Wise and Stadelman 1959) to be related at a highly significant level to the depth at which the samples were taken. Genetic makeup and, to a lesser extent, environmental factors such as feeding and management also influence tenderness.

The moisture content of chicken meat from fowl of roughly comparable history had little connection with tenderness (Dodge 1959). Furthermore, exercise and hormone treatments had little or no effect on tenderness of chicken muscle in one study.

Tenderness, juiciness, and flavor are all correlated in cooked pork loins according to Judge *et al.* (1960). Tenderness increased as the loins became less firm and as the pH decreased. Dark firm muscle was higher in pH and lower in free water than was light soft muscle.

Sausage

Sausages are of two general kinds, the high-moisture types which are made from a so-called emulsion composed of added water and the meat constituents (sometimes with added fat), and the low moisture types which (supposedly) do not contain added water and which may be partially dried by smoking or like processes. Both kinds are usually made of low grade meats having tough fibers. Texture in the wet type is to a large extent controlled by the amount of water added, the water-binding capacity of the proteins, and the proportion of fat. The dry type frequently contains pieces of meat large enough to be sensed separately. Particle size, water-binding capacity, inherent fiber toughness, fat content, and processing conditions all play roles in determining the texture of the product.

There seems to be no reference in the literature to specific objective techniques for determining the texture of sausage. Although shear methods would probably be unsatisfactory for use with these products, the grinder procedures could easily be adapted to the measurement of sausage consistency.

The water-binding capacity of meat can be increased by additives such as tetrasodium pyrophosphate, sodium acid pyrophosphate, hexametaphosphate, and sodium polyphosphate. Even sodium chloride can, under certain conditions, increase imbibition capacity. The phosphates also play a part in fat emulsification (Labie 1958). The "emulsion" of sausage consists of a matrix of protein and water encapsulating particles of fat. Fluid retention at 32°F. from sodium chloride and other neutral salt solutions depends on the degree of ion absorption, anions being retained preferentially. The amount of fluid retained corresponds well with anion absorption (Sherman 1961A, B, and C).

Grau and Hamm (1953) described a method for determining the water-binding capacity of meat. In essence, it involved pressing a small piece of muscle tissue between filter paper and measuring the weight lost by the meat. By using this method, Wismer-Pedersen (1959) showed that much variation occurred between different hogs in imbibition capacity of muscles.

Mealiness is a rather poorly defined texture defect which has been studied in connection with storage deterioration in dehydrated minced cooked pork (Bhatia 1959). Apparently no objective test method has been used to measure mealiness. It is evidently related to lowered imbibition capacity. Bhatia indicated that browning reactions are connected with the development of mealiness.

SUMMARY

Fibrous structures contribute significantly to the perceived texture of many vegetables but contribute to textural defects only in the mature stages of such products as asparagus, snap beans, celery, etc. Measurement of fibrousness is best accomplished by determining resistance to shear although methods involving disintegration or digestion of the tissues with subsequent separation and weighing of the fibers also give reliable results.

Panel testing is a very accurate means of determining the fibrousness of both vegetables and meats. The Warner-Bratzler shear apparatus and methods depending upon the measurement of the energy required to grind a sample are reasonably satisfactory objective procedures for estimating the texture of meat, including the fibrous component of the texture. It does not appear that measurements of the extensibility of single muscle fibers have a close enough relationship to subjective tenderness to enable the latter to be reliably predicted from the former. No chemical tests have been shown to yield results usefully correlated with fibrousness of meats.

BIBLIOGRAPHY

BACKINGER, G. T., KRAMER, A. K., DECKER, R. W., and SIDWELL, A. P. 1957. Application of work measurement to the determination of fibrousness in asparagus. Food Technol. 11, 583–585.

BATZER, O. F., SLIWINSKI, R. A., CHANG, L., PIH, K., FOX, J. B., JR., DOTY, D. M., PEARSON, A. M., and SPOONER, N. E. 1959. Some factors influencing radiation induced chemical changes in raw beef. Food Technol. 13, 501–508.

BHATIA, B. S. 1959. Role of products of non-enzymatic browning on development of mealiness in dehydrated cooked pork mince during storage in air. Food Sci., Mysore, 8, 309–312.

BIRNER, M. L., and AUERBACH, E. 1960. Microscopic structure of animal tissues. In The Science of Meat and Meat Products. W. H. Freeman and Co., San Francisco.

BOCKIAN, A. H., ANGLEMIER, A. F., and SATHER, L. A. 1958. A comparison of an objective and subjective measurement of beef tenderness. Food Technol. 12, 483–485.

BRAMBLETT, J. D., HOSTETLER, R. L., VAIL, G. E., and DRAUDT, H. N. 1959. Qualities of beef as affected by cooking at very low temperatures for long periods of time. Food Technol. 13, 707–711.

BRATZLER, L. J. 1932. Measuring the tenderness of meat by means of a mechanical shear. Master of Science Thesis, Kansas State College.

BRATZLER, L. J. 1949. Determining the tenderness of meat by use of the Warner-Bratzler method. Proc. Second Annual Reciprocal Meat Conference, National Livestock and Meat Board 1949.

BRAUNS, F. E., and BRAUNS, D. A. 1960. The Chemistry of Lignin. Academic Press, New York.

BRENNAN, J. R. 1959. Anatomical and respiration studies of the asparagus stem and the development of fibrousness during storage. Dissertation Abstracts 20, 40.

BRISKEY, E. J. 1961. Relationship of feeding to meat composition and properties. American Meat Institute Foundation Circ. 64, 35–46.

BROWN, S. A. 1961. Chemistry of lignification. Science 134, 305–313.

CALDWELL, J. S., and CULPEPPER, C. W. 1943A. Snap bean varieties suitable for dehydration. Part 2. Food Packer 24, 363–368.

CALDWELL, J. S., and CULPEPPER, C. W. 1943B. Snap bean varieties suitable for dehydration. Part 3. Food Packer 24, 420, 422, 424.

CALDWELL, J. S., CULPEPPER, C. W., HUTCHINS, M. C., EZELL, B. D., and WILCOX, M. S. 1944. Further studies of varietal suitability for dehydration in snap beans. Canner 99, No. 10, 12–15, 30.

CAMPBELL, G. W. 1956. Consumer acceptance of beef. A controlled retail store experiment. Ariz. Univ. Agr. Expt. Sta. Rept. 145.

CONNELL, J. J. 1957. Some aspects of the texture of dehydrated fish. J. Sci. Food Agr. 8, 526–537.

COVER, S., KING, G. T., and BUTLER, O. D. 1958. Effect of carcass grades and fatness on tenderness of meat from steers of known history. Texas Agr. Expt. Sta. Bull. 889.

CULPEPPER, C. W., and MAGOON, C. A. 1924. Studies upon the relative merits of sweet corn varieties for canning purposes and the relation of maturity of corn to quality of the canned product. J. Agr. Res. 28, 403–443.

CULPEPPER, C. W., and MOON, H. H. 1935. Composition of the developing asparagus shoot in relation to its use as a food product and as material for canning. U. S. Dept. Agr. Tech. Bull. 462.

DEATHERAGE, F. E., and GARNATZ, G. 1952. A comparative study of tenderness determination by sensory panel and by shear strength measurements. Food Technol. 6, 260–262.

DE FREMERY, D., and POOL, M. F. 1960. Biochemistry of chicken muscle as related to rigor mortis and tenderization. Food Research 25, 73–87.

DODGE, J. W. 1959. Factors involved with early post-mortem tenderness of poultry meat. Dissertation Abstracts 20, 1501–1502.

DODGE, J. W., and STADELMAN, W. J. 1959. Post mortem aging of poultry meat and its effect on the tenderness of the breast muscles. Food Technol. 13, 81–84.

DUNSING, M. 1959A. Visual and eating preferences of consumer household panel for beef animals of different ages. Food Technol. 13, 332–336.

DUNSING, M. 1959B. Visual and eating preferences of consumer household panel for beef from Brahma-Hereford crossbred and from Herefords. Food Technol. 13, 451–456.

EMERSON, J. A., and PALMER, A. Z. 1960. A food grinder-recording ammeter method for measuring beef tenderness. Food Technol. 14, 214–216.

EWELL, A. W. 1940. The tendering of beef. Refrig. Eng. 39, 237–240.

FUKAZAWA, T., HASHIMOTO, Y., and YASUI, T. 1961. Effect of storage on some physicochemical properties in experimental sausage prepared from fibrils. J. Food Sci. 26, 331–336.

GOULD, W. A. 1949. Instrument to quickly reveal quality of snap and wax beans. Food Packer 30, No. 12, 26–27.

GOULD, W. A. 1950A. Here's heat unit guide for 47 varieties of snap beans. Food Packer 31, No. 3, 35–37.

GOULD, W. A. 1950B. Seed length is good measure of maturity for snap beans. Food Packer 31, No. 8, 54–55.

GOULD, W. A. 1950C. What factors produce a fancy pack bean? Food Packer 31, No. 5, 26–27, 68, 70.

GOULD, W. A. 1951. Quality evaluation of fresh frozen, and canned snap beans. Ohio Agr. Expt. Sta. Bull. 701.

GRAU, R., and HAMM, R. 1953. A simple method for determining water binding in muscle. Naturwissenschaften 40, 29–30.

GUYER, R. B., and KRAMER, A. 1951. Studies of factors affecting the quality of green and wax beans. Maryland Agr. Expt. Sta. Bull. A68.

HAWKE, J. C. 1957. Consumer trials for the determination of the acceptability of dehydrated beef. J. Sci. Food Agr. 8, 197–205.

HINER, R. L., HARKINS, O. G., SLOANE, H. S., FELLERS, C. R., and ANDERSON, E. E. 1953. Fiber diameter in relation to tenderness of beef muscle. Food Research 18, 364–372.

HOWARD, A. 1956. Sensory tests of the quality of meat. Food Preserv. Quart. 16, 26–30.

HUSAINI, S. A., DEATHERAGE, F. E., and KUNKLE, L. E. 1950. Studies of meat. II. Observations on relation of biochemical factors to changes in tenderness. Food Technol. 4, 366–369.

HYUN, K. P. 1959. Factors influencing pithiness in the radish (Raphanus sativa L.). Dissertation Abstracts 20, 21.

ISHERWOOD, F. A. 1955. Texture in fruits and vegetables. Food Manuf. 30, 399–402, 420.

ISHERWOOD, F. A. 1960. Texture of plant tissues. Soc. Chem. Ind. Monograph 7, 135–143.

JUDGE, M. D., CAHILL, V. R., KUNKLE, L. E., and DEATHERAGE, F. E. 1960. Pork quality. II. Physical, chemical, and organoleptic relationships in fresh pork. J. Animal Sci. 19, 145–149.

KAUFFMAN, R. G., BRAY, R. W., and SCHAARS, M. A. 1961. Price vs. marbling in the purchase of pork chops. Food Technol. 15, 22–24.

KERTESZ, Z. I., EUCARE, M., and FOX, G. 1959. A study of apple cellulose. Food Research 24, 14–19.

KINNISON, A. F., and FINCH, A. H. 1933. Some effects of special practices influencing the nutritional balance on yield, texture, and time of maturity of grapefruit. Proc. Am. Soc. Hort. Sci. 30, 95–97.

KLIMCZAK, Z., and PIEKARSKA, J. 1957. Content and composition of cellulose in vegetables at different stages of growth. Roczniki Panstwowego Zakladu Hig. 8, 67–73.

KLOSE, A. A., HANSON, H. L., POOL, M. F., and LINEWEAVER, H. 1956. Poultry tenderness improved by holding before freezing. Quick Frozen Foods 18, No. 10, 95.

KRAMER, A. 1957. Food texture rapidly gauged with versatile shear-press. Food Eng. 29, No. 5, 57.

KRAMER, A., and AAMLID, K. 1953. The shear-press, an instrument for measuring the quality of foods. III. Application to peas. Proc. Am. Soc. Hort. Sci. 61, 417–423.

KRAMER, A., AAMLID, K., GUYER, R. B., and ROGERS, H. P., JR. 1951. New shear-press predicts quality of canned lima beans. Food Eng. 23, 112–113, 187.

KRAMER, A., and BACKINGER, G. 1959. Textural measurement of foods. Food 28, 85–86, 95.

KRAMER, A., BURKHARDT, G. J., and ROGERS, H. P. 1951. The shear press, an instrument for measuring quality of foods. I. The instrument. Canner 112, No. 5, 34–36, 40.

KRAMER, A., SCOTT, L. E., GUYER, R. B., and IDE, L. E. 1950. Factors affecting the objective and organoleptic evaluation of quality in raw and canned peas. Food Technol. 4, 142–150.

KRAMER, A., and SMITH, H. R. 1945. Reports to the labelling committee. National Canners Assoc. Mimeo. Rept.

KRAMER, A., SMITH, H. R. 1946. The succulometer. Canning Trade 68, No. 45, 7–8, 24.

KROPF, D. H., and GRAF, R. L. 1959A. Interrelationships of subjective, chemical, and sensory evaluations of beef quality. Food Technol. *13*, 492–495.

KROPF, D. H., and GRAF, R. L. 1959B. The effect of grade, weight, and class of beef carcasses upon certain chemical and sensory evaluations of beef quality. Food Technol. *13*, 719–721.

LABIE, C. 1958. The use of phosphates in the preservation of meat products. Rec. Med. Vet. Ecole d'Alfort *134*, 133–138.

LAW, N. H. 1955. Palatability of chilled and frozen beef. Food Manuf. *30*, 187–189.

LLOYD, E. J., and HINER, R. L. 1959. Relation between hydroxyproline of alkali-insoluble protein and tenderness of bovine muscle. J. Agr. Food Chem. 7, 860–862.

LOCKER, R. H. 1960. Degree of muscular contraction as a factor in the tenderness of beef. Food Research 25, 304–307.

LÖRINCZ, F., and SZEREDY, I. 1959. Quantitative and qualitative determination of connective tissue content of meat and meat products. J. Sci. Food Agr. *10*, 468–472.

LOVE, R. M. 1960. Texture change in fish and its measurement. Soc. Chem. Ind. Monograph 7, 109–118.

LYNCH, L. J., and MITCHELL, R. S. 1950. The physical measurement of quality in canning of peas. Commonwealth Sci. Ind. Research Organization Bull., Australia, *254*.

MACGILLIVRAY, J. H. 1933. Seasonal variation in tenderness of asparagus. Proc. Am. Soc. Hort. Sci. *30*, 558–560.

MACKINTOSH, D. L., HALL, J. L., and VAIL, G. E. 1936. Some observations pertaining to tenderness of meat. Proc. Am. Soc. Animal Production 29, 285–289.

MAHONEY, C. H. 1934. The relation of color and maturity of pods with tenderometer readings, maturity grade scores, and alcohol insoluble solids of peas grown for canning. Proc. Am. Soc. Hort. Sci. 37, 725–728.

MAKOWER, R. 1950. Methods of measuring the tenderness and maturity of processed peas. Food Technol. 4, 403–408.

MARTIN, W. M. 1937A. Apparatus for evaluating tenderness in peas. The Canner *84*, No. 12, 108–112.

MARTIN, W. M. 1937B. The tenderometer. An apparatus for evaluating tenderness in peas. Canning Trade *57*, No. 29, 7–8, 10, 12, 14.

MARTIN, W. M., LUECK, R. H., and SALLEE, E. D. 1938. Practical application of the tenderometer in grading peas. Canning Age *19*, 146–149.

MAY, K. N., SAFFLE, R. L., DOWNING, D. L., and POWERS, J. J. 1962. Interrelations of post-mortem changes with tenderness of chicken and pork. Food Technol. *16*, 72–73.

MCCONNELL, J. E. W. 1956. Factors affecting the skin of processed green beans. National Canners Assoc. Confidential Research Report *205-56*.

MIYADA, D. S., and TAPPEL, A. L. 1956. Meat tenderization. I. Two mechanical devices for measuring texture. Food Technol. *10*, 142–145.

MIYAUCHI, D. T. 1960. Irradiation preservation of Pacific Northwest fish. I. Cod fillets. Food Technol. *14*, 379–382.

MOYER, J. C., LYNCH, L. J., and MITCHELL, R. S. 1954. The tenderization of peas during vining. Food Technol. 8, 358–360.

NEUMANN, H. J., FRAME, L. R., MORGAN, L., and OLSON, R. L. 1961. Use of the Kramer shear-press in forecasting harvest dates for Fordhook lima beans. Food Technol. *15*, 225–228.

ORCHARD, J. E. 1961. Private communication.

PALMER, A. Z., CARPENTER, J. W., ALSMEYER, R. H., CHAPMAN, H. L., and KIRK, W. G. 1958. Simple correlation between carcass grade, marbling ether extract of loin eye, and beef tenderness. Animal Sci. 17, 1153.

PAUL, P. C. 1957. Tenderness of beef. J. Am. Dietetic Assoc. *33*, 890–894.

PAUL, P. C., BEAN, M., and BRATZLER, L. J. 1956. Effect of cold storage and method of cooking on commercial grade cow beef. Mich. Agr. Expt. Sta. Tech. Bull. *256*.

PAUL, P. C., SORENSON, C. I., and ABPLANALP, H. 1959. Variability in tenderness of chicken. Food Research 24, 205–209.

PETERSON, D. W., SIMONE, M. LILYBLADE, A. L., and MARTIN, R. 1959. Some factors affecting intensity of flavor and toughness of chicken muscle. Food Technol. 13, 204–207.

RAMSBOTTOM, J. M., STRANDINE, E. J., and KOONZ, C. H. 1945. Comparative tenderness of representative beef muscles. Food Research 10, 497–505.

REEVE, R. M. 1947. Relation of histological characteristics to texture in seed coats of peas. Food Research 12, 10–23.

REEVE, R. M. 1949. Histological observations on the seed coats of succulent peas. Food Research 14, 77–89.

RHODES, V. J., KIEHL, E. R., and BRADY, D. E. 1955. Visual preferences for grades of retail beef cuts. Missouri Agr. Expt. Sta. Res. Bull. 583.

ROWE, S. C., and BONNEY, V. R. 1936. A study of chemical methods for determining the maturity of canned snap beans. J. Assoc. Offic. Agr. Chemists 19, 620–628.

SAFFLE, R. L., and BRATZLER, L. J. 1959. The effect of fatness on some processing and palatability characteristics of pork carcasses. Food Technol. 13, 236–239.

SALE, A. J. H. 1960. Measurement of meat tenderness. Soc. Chem. Ind. Monograph 7, 103–108.

SAYRE, C. B. 1954. Comparison of the tenderometer and maturometer for measuring the quality of raw peas. Proc. Am. Hort. Sci. 63, 371–377.

SCHNEIDER, A. 1956. The determination of the qualitative properties of canned peas. Instr. Obst.-u. Gemuseverw. 41, 469–470.

SCHULTZ, H. W. 1957. An evaluation of the methods of measuring tenderness. Proc. Tenth Annual Reciprocal Meat Conference, National Livestock and Meat Board, 1957.

SHALLENBERGER, R. S., and MOYER, J. C. 1961. Sugar-starch transformations in peas. J. Agr. Food Chem. 9, 137–140.

SHERMAN, P. 1961A. The water binding capacity of fresh pork. I. The influence of sodium chloride, pyrophosphate, and polyphosphate on water absorption. Food Technol. 15, 79–87.

SHERMAN, P. 1961B. The water binding capacity of fresh pork. II. The influence of phosphates on fat distribution in meat products. Food Technol. 15, 87–89.

SHERMAN, P. 1961C. The water binding capacity of fresh pork. III. The influence of cooking temperature on the water binding capacity of lean pork. Food Technol. 15, 90–94.

SHEWFELT, A. L., and MOHR, W. P. 1960. Effect of spear length, spear portion, and harvest date on the composition and quality of asparagus. Can J. Plant Sci. 40, 371–374.

SIMONE, M., CARROLL, F., and CHICHESTER, C. O. 1959. Differences in eating quality factors of beef from 18- and 30-month steers. Food Technol. 13, 337–340.

SISTRUNK, W. A. 1959. Effect of certain field and processing factors on the texture of Blue Lake green beans. Dissertation Abstracts 20, 983–984.

SISTRUNK, W. A., and CAIN, R. F. 1960. Chemical and physical changes in green beans during preparation and canning. Food Technol. 14, 357–362.

SLEETH, R. B., HENRICKSON, R. L., and BRADY, D. E. 1957. Effect of controlling environmental conditions during aging on the quality of beef. Food Technol. 11, 205–208.

SLEETH, R. B., KELLEY, G. G., and BRADY, D. E. 1958. Shrinkage and organoleptic characteristics of beef aged in controlled environments. Food Technol. 12, 86–90.

SMITH, E. E. 1957. Tenderness and chemical changes in muscle fibers during cooking. Masters Thesis, Purdue University.

SPERRING, D. D., PLATT, W. T., and HINER, R. L. 1956. Tenderness in beef muscle as measured by pressure. Food Technol. 13, 155–158.

STARK, F. C., JR., and MAHONEY, C. H. 1942. A study of the development of the fibrous sheat in the side wall of edible snap bean pods with respect to quality. Proc. Am. Soc. Hort. Sci. 41, 351–359.

STERLING, C. 1954. Sclereid development and texture of Bartlett pears. Food Research *19*, 433–443.

SWEETMAN, M. D., and MACKELLAR, I. 1954. Food Selection and Preparation. John Wiley and Sons, New York.

SWIFT, C. E., WEIR, C. E., and HANKINS, O. G. 1954. The effect of variations in moisture and fat content on the juiciness and tenderness of bologna. Food Technol. *8*, 339–340.

TRESSLER, D. K., BIRDSEYE, C., and MURRAY, W. T. 1932. Tenderness of meat. I. Determination of relative tenderness of chilled and quick-frozen beef. Ind. Eng. Chem. *24*, 242–245.

VAN BUREN, J. P., MOYER, J. C., WILSON, D. E., ROBINSON, W. B., and HAND, D. B. 1960. Influence of blanching conditions on sloughing, splitting, and firmness of canned snap beans. Food Technol. *14*, 233–236.

VEIS, A. 1961. The structure and properties of collagen. American Meat Institute Foundation Circ. *64*, 95–105.

VISSER, R. Y., HARRISON, D. L., GOERTZ, G. E., BUNYAN, M., SKELTON, M. M., and MACKINTOSH, D. L. 1960. The effect of degree of doneness on the tenderness and juiciness of beef cooked in the oven and in deep fat. Food Technol. *14*, 193–198.

WALLS, E. P., and KEMP, W. B. 1939A. Preliminary report on the use of tenderometer in forecasting grades of Alaska peas for canning. Maryland Agr. Expt. Sta. Misc. Publ. *677*.

WALLS, E. P., and KEMP, W. B. 1939B. Relationship between tenderometer readings and alcohol insoluble solids of Alaska peas. Proc. Am. Soc. Hort. Sci. *37*, 729–730.

WEIER, T. E., and STOCKING, C. R. 1950. Histological changes induced in fruits and vegetables by processing. Advances in Food Research *2*, 297–342.

WEINBERG, B., and ROSE, D. 1960. Changes in protein extractability during post-rigor tenderization of chicken breast muscle. Food Technol. *14*, 376–378.

WILDER, H. K. 1948. Instructions for use of the fiberometer in the measurement of fiber content in canned asparagus. National Canners Assoc. Research Labs. Rept. *12313-C*.

WISE, R. G., and STADELMAN, W. J. 1959. Tenderness of various muscle depths associated with poultry processing techniques. Food Technol. *13*, 689–691.

WISMER-PEDERSEN, J. 1959. Quality of pork in relation to rate of pH change post mortem. Food Research *24*, 711–727.

CHAPTER 6

Texture Determinants in Cell Aggregates

INTRODUCTION

In liquid foods and in most manufactured foods the contribution of cell masses to texture is usually slight or non-existent. However, in fruit and vegetable products other than purées and juices, the physical properties of combinations of cells are often the chief determinants of textural characteristics. The importance of these properties is especially great when whole fruits and vegetables are being considered.

It is informative to divide into three categories the qualities of whole cells and cell agglomerates which affect perceived texture. These are: (1) the intercellular forces which bind the cells together, (2) the mechanical strength and rigidity due to structural elements and supporting members in the cell wall itself, and (3) the turgidity of the cell originating from the osmotic pressure of the cytoplasmic fluids.

Each of these categories defines a completely separate and distinct group of forces which contributes to the textural characteristics of certain foods. Although it is frequently possible to identify the compounds responsible for each of the effects, a determination of the relative contribution of each group of forces to physical measurements on the whole substances often cannot be made with any great degree of certainty.

In the following sections, each of these categories of properties is discussed in detail. The final section includes a discussion of methods for measuring texture in cell aggregates.

MECHANICAL PROPERTIES OF THE CELL WALL

Cell walls consist of a very heterogeneous collection of substances. Carbohydrates, proteins, lignin, and lipids saponifiable and non-saponifiable are the predominant classes of compounds which have been identified. In fruits and vegetables, polysaccharides are the main constituents in terms of quantity. Pectic substances, cellulose, and simple or complex polymers of xylose, galactose, arabose, and mannose are some of the carbohydrate substances present. On the other hand, it is not thought that amylose, amylopectin, or glycogen are present in the cell wall. Most of the investigators concerned with the contribution of the cell wall to food texture have tended to concentrate on examinations of the carbohydrate fractions.

Some of the texture alterations associated with ripening are thought to

104

be due to changes in the cell wall. Microscopic studies have shown that, in many cases, the cell walls actually become thinner and indicate loss of rigidity by rounding off or breaking as fruit matures. Addoms *et al.* (1930) observed these changes in Elberta peaches but found that they did not occur in the Shipper cling variety. Of the two, the former variety becomes much softer during ripening. Isherwood (1960) showed that a fall in total polysaccharide content which he observed in overripe Conference pears is accomplished by a drastic thinning of the cell walls.

On the other hand, the cell walls in vegetables such as peas and beans may actually become thicker as they mature. Reeve (1947 and 1949) has described structures in the skins of succulent peas which influence texture. The two outermost layers of the seed coat become highly specialized in wall structures. The outer layer, the macrosclereids, have cell walls which at maturity possess characteristic wall thickenings forming an internal fluted structure in the upper portions of each cell. Underlying the macrosclereids is a single layer of cells called oteosclereids which are hour-glass shaped and which also exhibit secondary wall thickenings at maturity. These changes are almost certainly due to fluctuations in the amount and type of carbohydrate in the cell walls.

Cellulose is present in significant amounts in the walls of all fruits and vegetables. A full elucidation of its role in texture has been hindered by the dearth of satisfactory methods for studying it in the native state. The effects of pectic compounds are better understood. They are present in cells both as cementing substances and as thickened regions in the walls. It is reasonable to assume that both forms of the substances contribute to the strength of the cell wall even though their major effect on texture is probably the result of the cementing function of the pectic materials in the middle lamellae.

Probably the most penetrating analysis of the status of cellulose in the cell walls of fruits is that of Sterling (1961). In a paper concerned with the physical state of cellulose during the ripening of peaches he summarized the findings of previous workers by saying "All available electron microscope evidence shows that cellulose in higher plants is organized exclusively in the form of rather dense microfibrils which are embedded in an unorganized matrix of other wall components." In this particular investigation, Sterling purified the cell walls of peaches of most of their non-cellulosic components and examined the residue by x-ray diffraction techniques. Definite increases in micellar diameter were observed as maturation progressed. There were probably increases in the degree of crystallinity. He interpreted these changes as evidences of a limited degradation of cellulose. Although he did not speculate on the effects of the changes on the overall texture of the fruits, it would seem that the increased crystallinity which he observed could decrease the mechanical strength of the cell wall. Kertesz *et al.* (1959) also examined the effect of cellulose on apple firmness. Their conclusions are summarized in Table 9.

The changes of the contents of several of the cell wall carbohydrates of pears during ripening were determined by Jermyn and Isherwood (1956). It appears from their data that the total polysaccharide content of the fruit decreased irregularly until an extremely overripe stage was reached, at which time a rather large increase was observed. The cellulose content remained about constant until the pears became almost ripe and then began to fall. A continual decline in galactan content was evident throughout the observation period while the mannan fluctuated in an un-

TABLE 9

RELATIONSHIP BETWEEN SOME PROPERTIES AND CONSTITUENTS OF APPLES[1]

	Coefficient of Correlation[2]
Firmness of (unstored) apples vs. per cent cellulose	0.796**
Firmness of (unstored) apples vs. per cent. A. I. S.	0.716**
Firmness of (unstored) apples vs. firmness change per month of storage	0.620**
Cellulose per cent vs. firmness change per month of storage	0.418
Cellulose per cent vs. "maximum days" storage life	0.600*

[1] From Kertesz *et al.* (1959).
[2] Significance at the 5 per cent level is indicated by *, significance at the 1 per cent level is indicated by **.

predictable manner. The xylan showed a general tendency to increase in quantity while the amount of araban at first declined and then rose as the pears became overripe. The investigators regard araban as the most labile constituent. It is difficult to evaluate in terms of mechanical strength the changes which Jermyn and Isherwood reported.

Isherwood (1955) has been a strong proponent of the theory that lignin plays a very important role in fruit and vegetable texture. Lignin is an amorphous compound (or series of compounds) of high molecular weight containing phenolic groups and methoxyl groups. Many workers consider it to be present as a coating or encrustation on the cellulose fibers. In this form it is regarded as adding to the strength and rigidity of the cell wall.

Isherwood considers lignin to be important in toughening the fibrous elements and the supporting tissues of the plant. It can prevent softening and disintegration of the tissues during cooking. He describes changes occurring in runner beans as they mature as an example of how lignin affects food texture. Immature beans snap readily with a clean break when bent, but older beans exhibit a ragged or stringy fracture due to a so-called parchment layer. This tissue consists principally of cellulose fibers lying at an angle to the main axis of the pod. In the younger beans this layer is less apparent because the adhesion between the fibers is much weaker and, when cooked, is readily broken down by the teeth.

Isherwood attributes these observed differences to increasing lignification of the fibers as the beans mature. The lignin binds the cellulose fibers together in a layer that is not disintegrated by boiling water. He considers that the extent of lignification is affected by growing conditions and that drying of the pod causes increases. It is noteworthy in this connection that Morris and Wood (1956) found that the texture of dried beans deteriorates if they are stored at moisture contents above about 13 per cent. This moisture level is probably too low to permit enzymatic changes or ordinary chemical reactions to occur in the more stable structural components of the bean, but the more reactive lignin might be induced to undergo further polymerization.

The reverse of these changes seem to occur in many fruits where maturation is accompanied by softening. Son (1960) showed by electron microscopy that the cohesion of the cell-wall microfibrils of peaches and pears decreased as the fruits ripened. It was not suggested that this change was due to any alteration in the amount or kind of lignin.

CELL TURGIDITY

The textural qualities resulting from cell wall rigidity and intercellular adhesion can probably best be summed up as "toughness" and "firmness," while "crispness" can perhaps be attributed in large part to the properties contributed to the food by cell turgor. Because of the inexactness of these words, they are not as useful as one might desire, but they do permit a crude verbal summation of the organoleptic characteristics having their origin in the cell properties discussed in this chapter.

The crispness of lettuce, watermelon, celery, and many other fresh fruits and vegetables is principally a function of the turgor of the cell. In many of these foods, the cell wall itself contains a relatively insignificant amount of rigid structural elements. Form and texture depend upon the presence of large amounts of rather dilute solutions enclosed in selectively permeable membranes. This effect can be seen best in some of the melons, where loss of selective permeability, through heat damage or other means, results in complete collapse of the tissue and disappearance of the characteristic texture. On the other hand, celery and the rind of watermelon, for example, lose crispness under these circumstances but retain a tough texture with most of the original volume. The cell walls in such foods contain enough rigid structural elements to maintain cell form in the absence of turgor.

Water can be lost through the semi-permeable membrane in the vapor phase with a consequent reduction in turgidity. The rate at which evaporation occurs is related to the ambient relative humidity. Numerous workers have reported on the greater wilting that occurs in drier atmospheres and the apparent preservative effect observed when crisp products such as lettuce are kept in water or in storage areas where the humidity is maintained at high levels. Somewhat equivalent results can be obtained by wrapping the product in films resistant to moisture vapor transfer, in which case the product adjusts its own atmosphere. This procedure is now being used extensively with lettuce.

In a study of possible methods for extending the storage life of fresh vegetables, Cook et al. (1958) found that cabbage retained quality better at 32°F. than at 38°F., remaining "fresh and turgid" for eight weeks at the lower temperature. Celery also retained crispness longer at 32°F. than at higher temperatures. A similar response of lettuce and tomatoes to storage conditions was observed. Packing the foods in polyethylene bags decreased wilting. A similar investigation, reported by Parsons (1959), showed that cabbage stored equally well at 32°F. and 38°F. but deteriorated at 45°F. When the relative humidity was 92 per cent, wilting occurred unless the cabbages were stored in crates with polyethylene linings.

Penetrometer tests taken at high temperatures generally yield lower results than those taken at lower temperatures. Hartmann (1924) found pears to be firmer at 51°F. than at 97°F., Hartmann and Bullis (1929) observed that resistance of sweet cherries to penetration was nearly 30 per cent less at 90° than at 32°, and Hawkins and Sands (1920) showed that strawberries, blackberries, raspberries, and cherries required more pressure to be punctured at low temperatures than at high. On the other hand, Haller (1941) found very little difference between the firmness of apples at 34°F. and 77°F. Hartmann attributed the increases in firmness (toughness?) which he observed at the lower temperatures to loss of turgidity. The wilting of lettuce usually observed at low temperatures was explained by Parsons and Wright (1956) as due to the difficulty of maintaining high humidities at these temperatures. When the product is protected from air of low relative humidity by polyethylene wrapping, low temperatures assist in maintaining crispness.

There are no publications containing data quantitatively relating the turgor pressure or hydrostatic pressure to measurements of textural characteristics. Weckel et al. (1959) measured "turgidity or firmness" by forcing two parallel wires 0.019 in. in diameter through a one mm. midsection of canned potato, but it is clear that their use of the word is not in accord with the definition of turgidity used here. Turgor pressure is frequently determined by immersing cells in liquid of varying osmotic pressure for a brief time and then observing the cells microscopically for signs of plasmolysis. The turgor pressure is considered to be approximately equivalent to the osmotic pressure which plasmolyses 50 per cent of the cells. It might be possible to secure meaningful data relating turgor pressure to textural characteristics by rapidly bringing pieces of the material to about 140° to 150°F., thus destroying the selective permeability of the cell membrane, and then comparing the resistance to penetration with that of other samples not heat treated. Of course, other changes occur in this range also, and measurements would have to be made rapidly in order to minimize the effect of enzymatic changes. In spite of the complicating factors, data secured by the suggested procedure should allow a worthwhile estimate to be made of the contribution of turgor pressure to perceived texture.

Nuts comprise a group of crisp foods about whose texture very little has been written. It is probable that crispness in nuts is due to cellular distension by the large amounts of oil which are present. Pecans, for example, contain less than five per cent water and more than 70 per cent oil. It is of little value to consider the osmotic pressures of such systems in relation to cell turgidity. Destruction of the semi-permeable membrane does not seem to destroy the crispness of nuts. The cell wall is probably not readily wetted by the oil even when the membrane is denatured and the cells may retain all or part of their internal pressure under such conditions.

INTERCELLULAR ADHESION

Although pectic substances in the cell wall affect the mechanical strength of the wall itself, their major contribution to perceived texture is a result of their function of joining the cells together. Pectic acid is undoubtedly the most significant cell-cementing agent in foodstuffs of plant origin, although lignins may also play a significant role in some cases. In animal tissues, hyaluronic acid-protein complexes probably have the same function, but much less is known about these substances.

Pectic substances are high molecular weight polymers of galacturonic acid, perhaps with associated carbohydrate groups such as arabinose, xylose, etc. The pectic substances within the cell are esterified to a high degree with methanol, but the substance in the middle lamellae is largely pectic acid, the form containing very few ester linkages. The free acid groups in pectic acid bind calcium ions rather firmly, probably in "bridge" bonds between carboxyl groups on adjacent molecules. As a result, solutions of pectic acid form firm gels when calcium is added.

Softening of fruits, whether it is the result of normal maturation or of spoilage processes, is often accompanied by changes in the characteristics of the pectic substances. Doesburg (1957) determined the pectin content, degree of esterification, and molecular weight of the pectic substances in apples at weekly intervals for some months before and after harvesting. There appeared to be no shortening of the chain-length of pectin molecules during ripening. The correlation between changes in calcium content and solubility of pectin, the changes in composition of the mixture of organic acids in the fruits, and some evidence of a change in pH of the cell walls during this period indicated to Doesburg that solubilization of pectin during ripening might be due to translocation of calcium in the cell walls relative to the pectic acid loci. Most previous workers have related maturation softening (at least, that part of the texture change due to a decrease in adhesive forces between the cells) to enzymatic degradation of the middle lamellae. Convincing evidence for this latter change has not been put forth.

Some workers have used chelating agents such as the Versenes and oxalate to solubilize the calcium normally bound by the pectinic acid. Personius and Sharp (1938) demonstrated that treatment of potato pieces with oxalate caused a decreased adhesion between the cells. Many studies have shown that the addition of calcium to potato preparations increases adhesion. Pyke and Johnson (1940), Rhodes and Davies (1945), and Scharschmidt (1954) found that addition of calcium to canned potatoes decreased "sloughing" (loss of particles from the surface layers). Similarly, Simon et al. (1953) found that addition of calcium chloride to potatoes prior to drying reduced sloughing and mushiness in the rehydrated product.

Calcium has been added to many products besides potatoes in an attempt to counteract the excessive softening which frequently results from the canning process. Kertesz et al. (1940), Siegel (1939), and many others have shown that traces of calcium chloride in canned tomatoes improve the texture, appearance, and drained weight. Loconti and Kertesz (1941) attributed this response to formation of calcium pectate.

Although Powers et al. (1961) found that the amounts of major pectic fractions were not correlated with either drained weight or firmness in canned pimientos, acidification with citric acid or the addition of various calcium salts significantly affected these properties and the two treatments were synergistic. Hoover (1960) has also reported on the use of calcium salts for firming canned green and red sweet bell peppers. Other representative studies which may be consulted are those of Hoogzand and Doesburg (1961) on canned cauliflower, blanched apple slices (Guadagni 1950), fresh red cherries (Whittenberger and Hills 1953), and pasteurized apricots (Joux 1957). It would be interesting to determine whether or not the addition of calcium to canned raspberries would reduce the separation of drupelets (leading to a mushy product with many broken berries) reported by Leinbach *et al.* (1951). This undesirable condition is also present in frozen raspberries and would seem to be due to changes which might be counteracted by increases in calcium ion concentration.

An increase in the concentration of the smaller monovalent cations such as sodium might be expected to displace some of the calcium, and this phenomenon may be the cause of the softening of tissues and lower

Courtesy M. Buch and U. S. Department of Agriculture

Fig. 18.   Effect of Bruising and Aging on the Texture of Individual Cherries

Left—Control, processed immediately after harvesting (drained weight = 68 per cent). Right—Cherries bruised twice by dropping from a 2 ft. height. allowed to stand 8 hr. in air at room temperature and then 16 hr. in ice water before being processed (drained weight = 74 per cent).

drained weight of tomatoes canned with added table salt (Siegel 1938).

In intact fruit, enzymatic action on pectic substances might be expected to be relatively slow. Damage to the fruit which destroys permeability barriers without denaturing the enzymes could induce or accelerate texture changes. These changes could conceivably be either increased firmness or increased softness, depending upon whether pectinesterase or polygalacturonase activity was predominant. The deterioration in texture observed when frozen foods are thawed was attributed by Almási

(1952) to the activity of pectic enzymes. He stated that the cell-wall lesions caused by ice crystal formation permitted pectinase and pectinesterase to escape from their *in situ* state and destroy the pectic substances present in the cell.

Texture changes in the opposite direction are observed in peas and cherries subjected to mechanical damage during or shortly after harvest. These foods seem to become firmer during a short post-harvest storage period if they are slightly bruised (LaBelle and Moyer 1960; Makower and Ward 1950; and Whittenberger and Hills 1953). This change is more pronounced when the cherries are stored at 34°F. than it is when

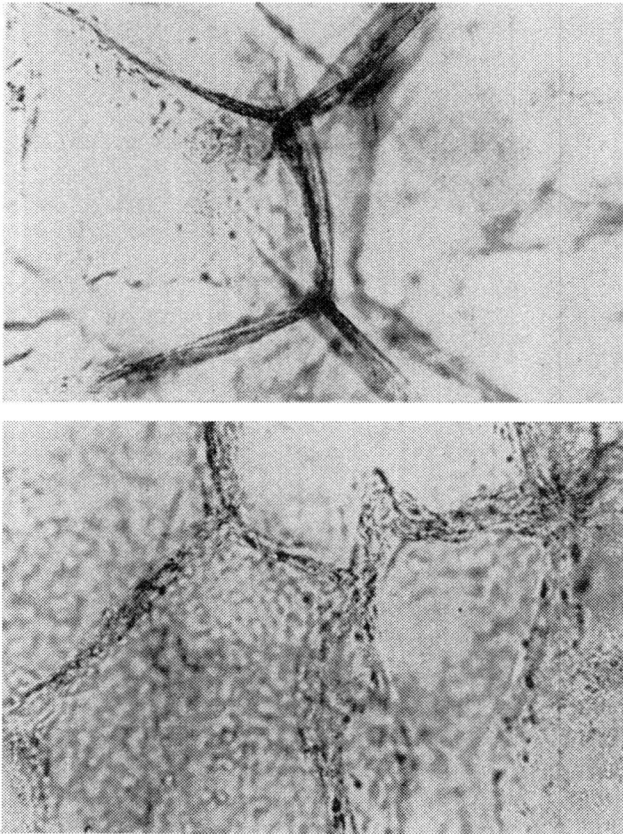

Courtesy M. Buch and U. S. Department of Agriculture

FIG. 19.   PHOTOMICROGRAPHS OF RADIAL SECTIONS OF TART RED CHERRIES AFTER EXTRACTION OF PECTIN WITH ACID, STAINED WITH METHYL GREEN

Control was canned immediately after harvesting. Firmed sample was allowed to stand 24 hr. at room temperature before being canned.

they are stored at 50° or 70°F. The cause of firming is not definitely known, but it is probably due to some hydrolysis of ester linkages in the more highly methoxylated compounds of the middle lamellae.

A recent study (Buch *et al.* 1961) definitely related the increased firmness of aged fresh cherries, which persists through the canning process, to an increased rigidity and adherence of the cell walls. Fig. 18 illustrates the difference in firmness resulting from bruising and aging. Such differences as are illustrated here can be detected readily in subjective tests. The microphotographs in Fig. 19 demonstrate the alterations in cell wall appearance which result from bruising. During the aging period, some compound is formed that imparts additional rigidity to the cell walls. Pectic acid may contribute to the increased rigidity and adhesion, but since, as Buch *et al.* showed, the rigidity is maintained even when all pectic acid is removed with sodium hydroxide, it is probable that another factor is involved.

The texture of cooked field-dried beans and peas appears to be related to the distribution of calcium in the product. According to Mattson (1946), the phytin and pectic acid in these legumes compete for the calcium ions. If insufficient phytin is present, the calcium ions will tend to saturate the carboxyl groups on the pectic acids causing unduly firm intercellular adhesion (Smithies 1960). Low phytin content can be the result of an initial insufficiency, or to breakdown by phytase during storage (Mattson *et al.* 1950).

As stated previously, the cementing function in animal tissue may be performed by hyaluronic acid-protein complexes. Hyaluronic acid is insolubilized by calcium ion, being similar in this respect to pectic acid. However, the relationship of calcium to the firmness of meat has been little studied. Carpenter *et al.* (1961) stated that muscle tissue infused post-rigor with the chelating agent sodium hexametaphosphate was more tender (as shown by taste panel and shear tests) than control muscles infused with tap water.

METHODS FOR MEASURING THE TEXTURE OF CELL AGGLOMERATES

Most methods used for evaluating the texture of pieces of food in their natural state and large enough to be perceived separately, fall into three main categories: (1) those which employ instruments to measure the force required to compress the whole fruit or vegetable (or large pieces of them); (2) those which rely upon the resistance to penetration of the food by one or more probes as an indication of texture, and (3) those which subject several pieces of the food to shearing forces. Some investigators have suggested the use of grinders electrically powered and connected (electrically or mechanically) with instruments capable of measuring the energy absorption, but these devices have not been widely adopted.

The use of spring-loaded probes to measure the texture or ripeness of fruits is quite old. O. M. Morris (1925) of the Washington Agricultural Experiment Station devised one of this type as early as 1917. This instrument, now important only from the historical point of view, consisted of a hard sphere (a marble) partly embedded in paraffin and resting on a spring scale. The whole fruit was pressed against the marble until the latter penetrated up to the paraffin and the pressure was then read off the scale. Although the deficiences of this method are rather obvious, variations of it are in use even today.

Lewis *et al.* (1919) described an instrument in which a cylindrical plunger activated by a lever was pressed into a fruit sample and the end point (a certain depth of penetration) was indicated by the closing of an electrical circuit. The Lewis device was the forerunner of an instrument designed by Magness and Taylor (1925) which has been used by many investigators and has served as the basis for several other pressure testers.

The Magness-Taylor apparatus is composed of a plunger surrounded by a metal collar and inclosed in a tubular handle. The plunger is connected to the barrel or handle by a calibrated steel spring. The spring is extended to an amount proportional to the force applied to the handle when the plunger tip is in contact with the test object. As the plunger is forced into the sample, the fruit surface pushes the collar into contact with an electrical junction. The light bulb which is energized by this contact determines the end point. Pressure is indicated on a scale attached to the barrel by a pointer which is part of the plunger. Magness and Taylor applied the device to apples after the skin of the fruit had been removed. The readings were used as an evaluation of the maturity of the fruit.

The Magness-Taylor pressure tester is still being used in one or another of its modifications. For example, Luh *et al.* (1959) used such an instrument with a $5/16$ inch plunger to determine the ripeness level in apricots. Kenworthy and Harris (1960) used a Magness-Taylor tester with a $7/16$ inch plunger to measure maturity of apples of different varieties and sources. Whittenberger (1951) compared results obtained on apple slices by the Magness-Taylor instrument with resistance to compression and tensile strength measurements (see Table 10).

Among the many plunger-type testers which have been described in the literature are the Culpepper and Magoon (1924) device which used a thin brass wire as the plunger, the "U. S. D. A. tester" (Haller 1941), the Universal Precision Penetrometer (Francis *et al.* 1959), and the Ballauf pressure tester (Leonard *et al.* 1957). Since these vary only in relatively minor respects and are rather similar in their principles of operation, they will not be reviewed in detail here.

TABLE 10

RESULTS OF VARIOUS TEXTURE TESTS APPLIED TO APPLES[1]

| Apple Samples[2] | Cooked Tissues[3] | | | Raw Tissues |
| | Compression[4] | | | |
	Per cent at Pressure of 25 Cm.	Standard Error	Tensile[4] Strength, Gm./Mm.	Magness-Taylor Test, Lbs.
Stayman Winesap	40	2.6	0.27	11.1
Red Delicious A	39	2.3	0.32	9.8
Jonathan A	34	2.8	0.57	10.0
Jonathan B	30	1.6	0.63	10.4
McIntosh	30	3.9	0.45	9.3
Stark	28	3.0	0.39	12.7
Red Delicious B	11	1.6	2.08	13.1
York Imperial	6	0.6	2.01	22.0

[1] From Whittenberger (1951).
[2] The data do not necessarily indicate the comparative firmness of average examples of the apple varieties.
[3] The coefficient of correlation between the compression and tensile strength of cooked tissues is -0.94. The correlation coefficient between the compression of cooked tissues and the Magness-Taylor test on raw tissues is -0.81.
[4] For the compression test of each variety, one specimen from each of seven apples was used. For the tensile test, two specimens from each of seven apples were used.

Plunger-type pressure testers are simple to use and give readings which are often found to be related to the overall texture of the product as it is determined organoleptically. However, they are of much less value when the fruit is not of homogeneous consistency. The plunger tends to penetrate between fibers and other regions of firmer texture. Multiple readings must usually be made to assure representative data. Culpepper and Caldwell found it necessary to make ten tests at each of three areas on each peach: 6 measurements were made on the cheeks in each area, 1 at each end, 1 at the suture, and 1 opposite the suture.

Texture devices based upon the principle of measuring the resistance to compression of whole fruits and vegetables (or large pieces of them) have not been widely used although there are some advantages inherent in such procedures. For example, many consumers judge the ripeness of fruits by pressing them so that whole fruit texture has a definite effect on the sales value of the product. The use of such tests would also tend to minimize the chance of error due to measurements on areas of atypical physical properties, which is an ever present danger in fibrous foods, fruits with stone cells in the flesh, etc.

The Kattan Firm-o-meter (Kattan 1957) and a commercial device based on the same principles (Asco firmness meter [Garrett *et al.* 1960]) crush whole tomatoes with an encircling chain which is constricted by a fixed weight for a given time. The decrease in tomato diameter is an indication of the fruit's firmness. According to Garrett *et al.*, the Asco firmness meter measures attributes similar to those estimated by squeezing the fruit by hand. Results correlated at a highly significant level with panel rankings of sample firmness.

Since mastication is essentially a combination of crushing and grinding motions, it would appear that measurements of the resistance offered by a food to crushing or compression would be directly related to some components of the sensory texture. Two interesting devices which were designed to duplicate some of the chewing motions and forces are the denture tenderometer and its modification, the strain gage pea tenderometer (Davison *et al.* 1959). In these machines, peas (or, presumably, other products of small size) can be ground between the surfaces of commercial flat-surfaced denture molars made of steel. The resistance to "chewing" causes a change in the strain on the main drive shaft of the instrument. These changes are electrically sensed and recorded. The grinding action of the molar surfaces is arranged so as to simulate some of the chewing actions in the human mouth. The upper molar exerts a downward crushing motion and the lower molar exerts a two mm. lateral shearing motion. A limitation of this machine is that the size of the test object affects the results so that peas of uniform size must be selected to achieve statistically significant results. Since size, maturity, and texture are all interrelated, this restriction would appear to severely limit the utility of the device.

A crushing test was used by Whittenberger and Marshall (1950) to estimate the firmness of fresh red cherries. Each of 20 cherries was compressed for ten seconds between two flat discs 2 cm. in diameter. A force of approximately 300 gm. was applied for about ten seconds and the per cent compression measured. Here also, size was a factor affecting the results, as stated by the authors. A large cherry tended to be compressed more than a small one of similar inherent firmness of flesh.

Verner (1931) had earlier described a pressure tester for stone fruits in which the fruit was squeezed between two flat surfaces or discs for a given distance ("about" one-half inch). The lower surface rested on the platform of a spring scale. The top disc could be adjusted by a screw until it just touched the fruit and the rod to which it was attached was pivoted to reduce the shearing effect. This tester was initially used on sweet cherries and Italian prunes. Verner found it to be of questionable value in determining the picking times. Haller *et al.* (1933) used a Verner instrument in which the plunger moved through a distance of seven-eighths of an inch for determining the texture of strawberries.

The Delaware jelly-strength tester (described in the chapter on gels) was adapted by Whittenberger (1951) to measure the resistance of apples to crushing. Cylindrical samples cut from raw apples were evacuated, cooked, and then compressed. Samples which were equally firm as indicated by per cent compression, could be either tough or tender, as determined organoleptically. However, the tough samples tended to remain completely coherent upon compression and to recover when the stress was removed while the tender samples tended to collapse completely, i.e., the cells separated from each other and the tissues lost their continuity. Whittenberger found a correlation coefficient of -0.81 between the results obtained with Magness-Taylor tester on raw samples and those obtained with the pressure tester.

Powrie and Asselbergs (1957) used a very similar test on canned syrup-pack carrots. They cut from the carrots cylinders 15 mm. in diameter and 11.8 mm. long, making sure that the cylinders contained about equal amounts of phloem and xylem. The samples were compressed in the Delaware jelly strength tester using a force equivalent to 16 inches of carbon tetrachloride attained in 20 seconds. The decrease in cylinder height was recorded as an indication of firmness. Glegg et al. (1956) reported on the firmness of apples, carrots, and beets in terms of the load required to crush a cylindrical sample.

Test instruments which measure the resistance to shear by multiple blades of a collection of several individual food pieces, or large portions of the bigger fruits and vegetables, have the great practical advantage that the measurements obtained are based on an "average" texture. The chances that the results will be truly representative of the entire batch are much greater than would otherwise be the case. Furthermore, the entire fruit or vegetable is evaluated, not just the skin or parts of the flesh or other individual tissues. The recent tendency seems to be toward more use of instruments of this type in technological investigations of food texture.

A representative shearing instrument is the Lee Comptroller, a device apparently based on a long series of investigations and mechanical improvements by Amihud Kramer and his co-workers (Kramer and Backinger 1959). The test cells can be changed to suit the product being examined but in general they consist of an upper element composed of several blades or many square rods, and a lower rectangular container which serves to restrain the product. The container may have a solid bottom or the bottom may be perforated to allow passage of the penetrating blades. The top element is moved vertically by a hydraulic system while the container remains stationary. Interposed between the piston of the hydraulic press and the blades is a proving ring dynamometer, into which is fitted the measuring element, a transducer or a gage. When the former measuring element is used, electronic equipment can be used to chart the pounds of force versus the distance travelled through the sample. This instrument seems likely to have wide application due to its versatility and precision.

Wolodkewitsch (1956) described a shear press suitable for testing the firmness of stone fruits. Grünewald (1959) modified the device for testing peas. It consisted of four parts: (1) an upper part holding 55 dies

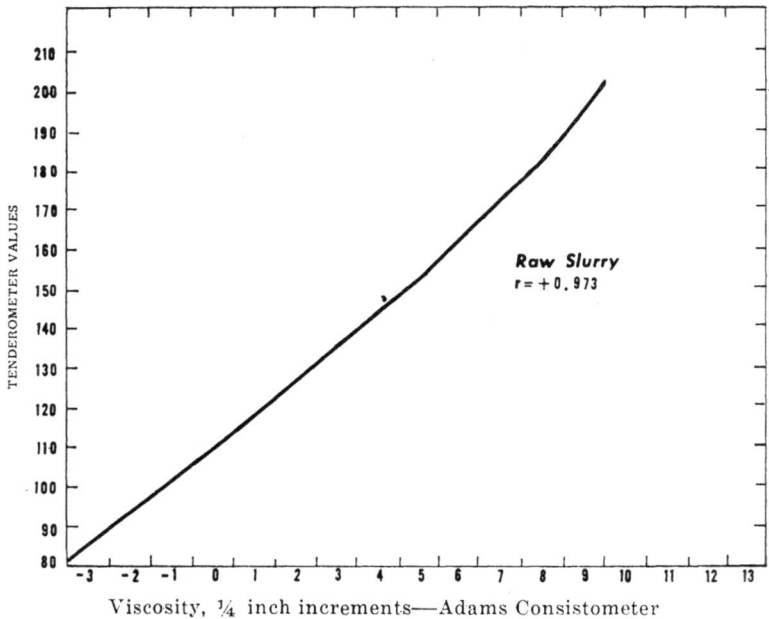

Viscosity, ¼ inch increments—Adams Consistometer

Fig. 20.   Relationship Between Raw Sweet Pea Slurry Viscosity and Tenderometer Values of the Whole Peas

each with a diameter of four mm., (2) a cover with 55 holes corresponding in arrangement to the dies and slightly larger, (3) a test container permitting 55 peas to be held in holes having a diameter of 4.4 mm., and (4) a container to catch the residue of the peas which is forced through the holes.

An instrument not fitting into any of the three suggested categories was designed by Massey et al. (1961). This device measured the extent of flexure or bending of lettuce leaves on which weights were applied.

When the data were corrected for leaf thickness, they were found to be related to crispness and other textural qualities. The water content of samples was found to be a major factor in crispness so it was necessary to equilibrate the leaves before testing by soaking them for two hours in 77°F. water.

It would not be expected that the rheological properties of ground fruits or vegetables would be of much value for predicting the firmness of the whole product. Therefore it is rather surprising that Wiley (1959) found a reliable relationship between the apparent viscosities of raw sweet peas disintegrated in a Waring blendor and the tenderometer values for the whole vegetable. This relationship is summarized by Fig. 20. Similar findings were reported for lima beans.

Various chemical tests have been suggested as means for predicting the texture of foods or physical characteristics related to texture. The alcohol-insoluble-solids content definitely varies with the stage of maturity in many fruits and vegetables and so is an indirect measure of texture in these products; for example see Torafson *et al.* (1956). As typified by the results of Isherwood reviewed in the preceding section, variations in specific carbohydrate fractions are too erratic to permit useful predictions of texture to be based on measurements of the content of any of these compounds. In corn and in some legumes, the moisture content is a reliable indicator of texture since the per cent dry matter tends to increase as the product becomes more mature and tougher. However, it is desirable to supplement the moisture test with such determinations as pericarp content and kernel size (Kramer 1952).

SUMMARY

The special properties of whole cells and cell agglomerates which affect food texture can be divided into three categories: (1) adhesion of cells to one another, (2) cell turgidity, and (3) rigidity of the cell walls.

Cohesion of plant cells is due principally to the pectic acid in the middle lamellae. The gel strength of pectic acid solutions is related to the molecular weight and to the concentration of calcium present (other cations also affect gel strength, sometimes adversely). It might be expected that the strength of intercellular bonding would also be related to the molecular size distribution and to the calcium content of the pectic substances in the middle lamellae. There is a great deal of indirect evidence that such is indeed the case.

Although decreases in the small amounts of ester groups which remain in pectic acid preparations do not greatly increase their gel strength, pectinesterase can function to strengthen intercellular forces by acting on pectins or pectinic acids associated with the pectic acids in the middle lamellae.

Cell turgor in the proper sense is found only in cells whose membranes possess selective permeability, but a certain amount of cellular distension can exist in other cells as a result of the presence of hydrated particles (such as gelatinized starch granules) or of solute molecules too large to diffuse through the discontinuities in the denatured membrane. In many fresh foods, such as lettuce and watermelon, the predominant textural characteristic is a crispness which is due almost entirely to cell turgor pressure.

The structural elements contributing to the rigidity and other mechanical properties of plant cell walls are cellulose fibers, pectin gels, and lignins. In fruits, the most important texture-influencing changes during maturation are alterations of the pectic substances and of the cell-wall cellulose which tend to make the fruit softer. These same changes occur in vegetables, but they are frequently overshadowed by the development of coarse fibrous elements and tough layered structures so that vegetables frequently tend to get tougher with increasing maturity.

There has been little or no publication of work concerned with measurement of the separate properties discussed in this chapter. Devices for measuring the texture of pieces of fruits and vegetables yield composite data in which the contributions of the separate properties are weighted by some unknown factor. Most of the instruments which have achieved wide usage by workers in the field depend upon one or more of the following principles: (1) measurement of the force required to compress the whole food or large pieces of it, (2) measurement of the resistance of the food to penetration by probes of rather small cross section, or (3) measurement of the force required to shear one or more pieces of the food.

BIBLIOGRAPHY

Addoms, R. M., Nightingale, G. T., and Blake, M. A. 1930. Development and ripening of peaches as correlated with physical characteristics, chemical composition, and histological structure of the fruit flesh. II. Histology and microchemistry. New Jersey Agr. Expt. Sta. Bull. 507.

Allen, F. W. 1932A. The harvesting and handling of fall and winter pears. Calif. Univ. Agr. Expt. Sta. Bull. 533.

Allen, F. W. 1932B. Physical and chemical changes in the ripening of deciduous fruits. Hilgardia 6, 381–441.

Almási, E. 1952. Investigations on the permeability of cells. Elelmezési Ipar 6, 82–86.

Appleman, C. O., and Conrad, C. M. 1926. Pectic constituents of peaches and their relation to softening of the fruit. Maryland Univ. Agr. Expt. Sta. Bull. 238.

Bailey, D. M., and Bailey, R. M. 1938. The relation of the pericarp to tenderness in sweet corn. Proc. Am. Soc. Hort. Sci. 36, 555–559.

BELL, T. A., ETCHELLS, J. L., and JONES, I. D. 1955. A method for testing cucumber salt-stock brine for softening activity. U. S. Dept. Agr., Agr. Res. Serv. ARS 72–5.

BIRKNER, M. L., and AUERBACH, E. 1960. Microscopic structure of animal tissues. In The Science of Meat and Meat Products. W. H. Freeman and Co., San Francisco, Calif.

BLAKE, M. A. 1929. A device for determining the texture of peach fruits for shipping and marketing. New Jersey Agr. Expt. Sta. Circ. 212.

BLAKE, M. A., DAVIDSON, O. W., ADDOMS, R. M., and NIGHTINGALE, G. T. 1931. Development and ripening of peaches as correlated with physical characteristics, chemical composition, and histological structure of the fruit flesh. I. Physical measurements of growth and flesh texture in relation to the market and edible qualities of the fruit. New Jersey Agr. Expt. Sta. Bull. 525

BONNEY, V. G., CLIFFORD, P. A., and LEPPER, H. A. 1931. An apparatus for determining the tenderness of certain canned fruits and vegetables. U. S. Dept. Agr. Circ. 164.

BUCH, M. L., SATORI, K. G., and HILLS, C. H. 1961. The effect of bruising and aging on the texture and pectic constituents of canned red tart cherries. Food Technol. 15, 526–531.

BULLIS, D. E., and WIEGAND, E. H. 1931. Bleaching and dyeing Royal Anne cherries for maraschino or fruit salad use. Oregon State Coll. Agr. Expt. Sta. Bull. 275.

CARPENTER, J. A., SAFFLE, R. L., and KAMSTRA, L. D. 1961. Tenderization of beef by post-rigor infusion of a chelating agent. Food Technol. 15, 197–199.

CARTER, G. H., INGALSBE, D. W., NEUBERT, A. M., and PROEBSTING, E. L., JR. 1958. Canning quality of Elberta peaches as affected by nitrogen fertilization. Food Technol. 12, 174–179.

CLAYPOOL, L. L., LEONARD, S., LUH, B. S., and SIMONE, M. 1958. Influence of ripening temperature, ripeness level, and growing area on quality of canned Bartlett pears. Food Technol. 12, 375–380.

COOK, H. T., PARSONS, C. S., and McCOLLOCH, L. P. 1958. Methods to extend storage of fresh vegetables aboard ships of the U. S. Navy. Food Technol. 12, 548–550.

CULPEPPER, C. W. 1936. Effect of stage of maturity of the snap bean on its composition and use as a food product. Food Research 1, 357–376.

CULPEPPER, C. W. 1937. Composition of summer squash and its relationship to variety, stage of maturity, and use as a food product. Food Research 2, 289–303.

CULPEPPER, C. W., and CALDWELL, J. S. 1930. The canning quality of certain commercially important Eastern peaches. U. S. Dept. Agr. Tech. Bull. 196.

CULPEPPER, C. W., CALDWELL, J. S., and MOON, H. H. 1935. A physiological study of development and ripening in the strawberry. J. Agr. Research 50, 645–696.

CULPEPPER, C. W., and MAGOON, C. A. 1924. Studies upon the relative merits of sweet corn varieties for canning purposes and the relation of maturity of corn to quality of the canned product. J. Agr. Research 28, 403–443.

DAVISON, S., BRODY, A. L., PROCTOR, B. E., and FELSENTHAL, P. 1959. A strain gage pea tenderometer. II. Instruments description and evaluation. Food Technol. 13, 119–123.

DEMAIN, A. L., and PHAFF, H. J. 1957. Cucumber curing: softening of cucumbers during curing. J. Agr. Food Chem. 5, 60–63.

DOESBURG, J. J. 1957. Relation between the solubilization of pectin and the fate of organic acids during maturation of apples. J. Sci. Food Agr. 8, 206–216.

ERNEST, J. V., BIRTH, G. S., SIDEWELL, A. P., and GOLUMBIC, C. 1958. Evaluation of light transmittance techniques for maturity measurements of two varieties of prune-type plums. Food Technol. 12, 595–599.

ETCHELLS, J. L., BELL, T. A., and WILLIAMS, C. F. 1958. Inhibition of pectinolytic and cellulolytic enzymes in cucumber fermentations by Scuppernong grape leaves. Food Technol. 12, 204–208.

122   FOOD TEXTURE

Francis, F. J., Amla, B. L., and Kiratsous, A. 1959. Control of exudation in pre-peeled French-fry potatoes with antibiotics. Food Technol. *13*, 485–488.

Garrett, A. W., Desrosier, N. W., Kuhn, G. D., and Fields, M. L. 1960. Evaluation of instruments to measure firmness of tomatoes. Food Technol. *14*, 562–564.

Glegg, R. E., Boyle, F. P., Tuttle, L. W., Wilson, D. E., and Kertesz, Z. I. 1956. Effects of ionizing radiations on plant tissues. I. Quantitative measurements of the softening of apples, beets, and carrot. Radiation Research *5*, 127–133.

Graham, R. W., and Evans, G. 1957. Standardization and operation of tenderometers. Food Manuf. *32*, 224–229.

Griswold, R. M. 1944. Factors influencing the quality of home-canned Montmorency cherries. Mich. State Univ. Agr. Expt. Sta. Tech. Bull. *194.*

Grünewald, T. 1959. Test apparatus for measuring the firmness of peas according to N. Wolodkewitsch. Z. Lebensm. Untersuch. u. Forsch. *110*, 93–94.

Guadagni, D. G. 1950. Adaptation of the tenderometer for the quantitative determination of firmness in calcium-treated apple slices. Food Technol. *4*, 319–321.

Hall, R. C., and Fryer, H. C. 1953. Consistency evaluation of dehydrated potato granules and directions for microscopic rupture count procedure. Food Technol. *7*, 373–377.

Haller, M. H. 1941. Fruit pressure testers and their practical applications. U. S. Dept. Agr. Circ. *627.*

Haller, M. H. 1952. Handling, transportation, storage, and marketing of peaches. U. S. Dept. Agr. Biblio. Bull. *21.*

Haller, M. H., and Harding, P. L. 1938. Relation of soil moisture to firmness and storage quality of apples. Proc. Am. Soc. Hort. Sci. *35*, 205–211.

Haller, M. H., Harding, P. L., and Rose, D. H. 1933. The interrelation of firmness, dry weight, and respiration in strawberries. Proc. Am. Soc. Hort. Sci. *29*, 330–334.

Hamilton, I. R., and Johnston, R. A. 1961A. Studies of cucumber softening under commercial salt-stock conditions in Ontario. I. Incidence and pattern of activity of pectolytic enzymes. Appl. Microbiol. *9*, 121–127.

Hamilton, I. R., and Johnston, R. A. 1961B. Studies of cucumber softening under commercial salt-stock conditions in Ontario. II. Pectolytic micro-organisms isolated. Appl. Microbiol. *9*, 128–123.

Hartmann, H. 1924. Studies relating to the harvesting and storing of apples and pears. Oregon State Coll. Agr. Expt. Sta. Bull. *206.*

Hartmann, H. 1926. Studies relating to the harvesting of Italian prunes for canning and fresh fruit shipment. Oregon State Coll. Agr. Expt. Sta. Circ. *75.*

Hartmann, H., and Bullis, D. E. 1929. Investigations relating to the handling of sweet cherries with special reference to chemical and physiological activities during ripening. Oregon State Coll. Agr. Expt. Sta. Bull. *247.*

Hawkins, L. A., and Sands, C. E. 1920. Effect of temperature on the resistance to wounding of certain small fruits and cherries. U. S. Dept. Agr. Bull. *830.*

Heinze, P. H., Kirkpatrick, M. E., and Dochterman, E. F. 1955. Cooking quality and compositional factors of potatoes of different varieties from several commercial locations. U. S. Dept. Agr. Tech. Bull. *1106.*

Hills, C. H., Whittenberger, R. T., Robertson, W. F., and Case, W. H. 1953. Studies on the processing of red cherries. II. Some effects of bruising on the yield and quality of canned Montmorency cherries. Food Technol. *7*, 32–35.

Holfelder, E. 1957. Experience in the use of the texturometer for measurements of the consistency of raw sauerkraut. Ind. Obst.-u. Gemüseverwert. *42*, 39–41.

Hoogzand, C., and Doesburg, J. J. 1961. Effect of blanching on texture and pectin of canned cauliflower. Food Technol. *15*, 160–163.

Hoover, M. W. 1960. Use of calcium hydroxide for firming canned green and red sweet bell peppers. Food Technol. *14*, 437–440.

Hough, L. F., and Weaver, G. M. 1959. Irradiation as an aid in fruit variety improvement. I. Mutations in the peach. J. Heredity *50*, 59–62.

ISHERWOOD, F. A. 1953. Texture in fruits and vegetables. Food Manuf. *30*, 399–402, 420.

ISHERWOOD, F. A. 1960. Texture of plant tissues. Soc. Chem. Ind. Monograph *7*, 135–143.

JERMYN, M. A., and ISHERWOOD, F. A. 1956. Changes in the cell-wall of the pear during ripening. Biochem. J. *64*, 123–132.

JOUX, J. L. 1957. Role of pectic substances in the maintenance of firmness in pasteurized apricots. Compt. rend. acad. agr. France *43*, 506–513.

KATTAN, A. A. 1957. Changes in color and firmness during ripening of detached tomatoes, and the use of a new instrument for measuring firmness. Proc. Am. Soc. Hort. Sci. *70*, 379–386.

KENWORTHY, A. L., and HARRIS, N. 1960. Organic acids in the apple as related to variety and source. Food Technol. *14*, 372–375.

KERTESZ, Z. I. 1951. The Pectic Substances. Interscience Publishers, New York.

KERTESZ, Z. I., EUCARE, M., and FOX, G. 1959. A study of apple cellulose. Food Research *24*, 14–19.

KERTESZ, Z. I., TOLMAN, T. G., LOCONTI, J. D., and RUYLE, E. H. 1940. The use of calcium in the commercial canning of whole tomatoes. N. Y. State Agr. Expt. Sta. Geneva, N. Y. Bull. *252*.

KIMBROUGH, W. D. 1931. The quality of strawberries as influenced by rainfall, soil moisture, and fertilizer treatments. Proc. Am. Soc. Hort. Sci. *27*, 184–186.

KRAMER, A. 1952. A tri-metric test for sweet corn quality. Proc. Am. Soc. Hort. Sci. *59*, 405–413.

KRAMER, A. and BACKINGER, G. 1959. Textural measurement of foods. Food *28*, 85–86, 95.

LABELLE, R. L., and MOYER, J. C. 1960. Factors affecting the drained weight and firmness of red tart cherries. Food Technol. *14*, 347–352.

LEINBACH, L. R., SEEGMILLER, C. G., and WILBUR, J. S. 1951. Composition of red raspberries, including pectin characterization. Food Technol. *5*, 51–54.

LEONARD, S., LUH, B. S., HINREINER, E., and SIMONE, M. 1957. Bartlett pears for canning. Calif. Agr. *11*, No. 2, 10, 15.

LEONARD, S., LUH, B. S., and MRAK, E. M. 1958. Factors influencing drained weight of canned clingstone peaches. Food Technol. *12*, 80–85.

LEWIS, C. I., MURNEEK, A. E., CATE, C. C. 1919. Pear harvesting and storage investigations in Rogue River Valley (Second report). Oregon State Coll. Agr. Expt. Sta. Bull. *162*.

LOCONTI, J. D., and KERTESZ, Z. I. 1941. Identification of calcium pectate as the tissue firming compound formed by treatment of tomatoes with calcium chloride. Food Research *6*, 499–505.

LUGT, C., and VEENBAAS, A. 1954. Comparison of the tenderometer and "Hardness meter" for measuring ripeness of peas. Conserva *2*, 367–371.

LUH, B. S., LEONARD, S. J., and MRAK, E. M. 1959. Drained weight of canned apricots. Food Technol. *13*, 253–257.

MAGNESS, J. R., DIEHL, H. C., and ALLEN, F. W. 1929. Investigations on the handling of Bartlett pears from Pacific coast districts. U. S. Dept. Agr. Tech. Bull. *140*.

MAGNESS, J. R., and TAYLOR, G. F. 1925. An improved type of pressure tester for the determination of fruit maturity. U. S. Dept. Agr. Dept. Circ. *350*.

MAKOWER, R. U., BOGGS, M. M., BURR, H. K., and OLCOTT, H. S. 1953. Comparison of methods for measuring the maturity factor in frozen peas. Food Technol. *7*, 43–48.

MAKOWER, R. U., and WARD, A. C. 1950. Role of bruising and delay in the development of off-flavor in peas. Food Technol. *14*, 47–49.

MASSEY, L. M., JR., TALLMAN, D. F., and KERTESZ, Z. I. 1961. Effects of ionizing radiations on plant tissues. V. Some effects of gamma radiation on lettuce leaves. J. Food Sci. *26*, 389–396.

MATTSON, S. 1946. The cookability of yellow peas. A colloid-chemical and biochemical study. Acta Agr. Suecana 2, 185–231.

MATTSON, S., ÅKERBERG, E., ERICKSSON, E., KOUTLER-ANDERSSON, E., and VAHTRAS, K. 1950. Factors determining the composition and cookability of peas. Acta Agr. Scand. 1, 40–61.

MITCHELL, R. S., CASIMIR, D. J., and LYNCH, L. J. 1961. The maturometer—Instrumental test and redesign. Food Technol. 15, 415–418.

MORRIS, H. J., and WOOD, E. R. 1956. Influence of moisture content on keeping quality of dry beans. Food Technol. 10, 225–229.

MORRIS, O. M. 1925. Studies in apple storage. Wash. State Coll. Agr. Expt. Sta. Bull. 193.

MURNEEK, A. E. 1921. A new test for maturity of the pear. Oregon State Coll. Agr. Expt. Sta. Bull. 186.

NICHOLAS, R. C., and PFLUG, I. J. 1960. Effects of high temperature storage on the quality of fresh cucumber pickles. Glass Packer 39, No. 4, 35, 38–39, 65.

PANGBORN, R. M., LEONARD, S., SIMONE, M., and LUH, B. S. 1959. Freestone peaches. I. Effect of sucrose, citric acid, and corn syrup on consumer acceptance. Food Technol. 13, 444–447.

PANGBORN, R. M., VAUGHN, R. H., YORK, G. K., III, and ESTELLE, M. 1959. Effect of sugar, storage time, and temperature on dill pickle quality. Food Technol. 13, 489–492.

PARSONS, C. S. 1959. Effects of temperature and packaging on quality of stored cabbage. Proc. Am. Soc. Hort. Sci. 74, 616–621.

PARSONS, C. S., and WRIGHT, R. C. 1956. Effects of temperature, trimming, and packaging methods on lettuce deterioration. Proc. Am. Soc. Hort. Sci. 68, 283–287.

PATTON, M. B., GORRELL, F. L., and BROWN, H. D. 1943. Relation of fertility levels to tenderness of garden beets. Proc. Am. Soc. Hort. Sci. 43, 225–228.

PERSONIUS, C. J., and SHARP, P. F. 1938. Adhesion of potato tissue cells as influenced by pectic solvents and precipitants. Food Research 4, 299–306.

POWERS, J. J., PRATT, D. E., DOWNING, D. L., and POWERS, I. T. 1961. Effect of acid level, calcium salts, monosodium glutamate, and sugar on canned pimientos. Food Technol. 15, 67–74.

POWRIE, W. D., and ASSELBERGS, E. A. 1957. A study of canned syrup-pack whole carrots. Food Technol. 11, 275–277.

PYKE, W. E., and JOHNSON, G. 1940. The relation of the calcium ion to the sloughing of potatoes. Am. Potato J. 17, 1–9.

REEVE, R. M. 1947. Relation of histological characteristics to texture in seed coats of peas. Food Research 12, 10–23.

RHODES, W. E., and DAVIES, H. F. 1945. The selection and pre-processing of potatoes for canning with special reference to control of texture by calcium chloride. Chem. Ind. London 21, 162–163.

ROBINSON, W. B., MOYER, J. C., and KERTESZ, Z. I. 1949. "Thermal maceration" of plant tissue. Plant Physiol. 24, 317–319.

ROOD, P. 1957. Development and evaluation of objective maturity indices for California freestone peaches. Proc. Am. Soc. Hort. Sci. 70, 104–112.

ROSE, D. H., HALLER, M. H., and HARDING, P. L. 1935. Relation of temperature of fruit to firmness in strawberries. Proc. Am. Soc. Hort. Sci. 32, 429–430.

ROSS, E. A. 1949. A quantitative hardness tester for food products. Science 109, 204.

ROSS, E., and SHOUP, N. H. 1950. "Hard-end" detected early. Food Packer 31, No. 7, 33, 36, 38.

RYALL, A. L., and ALDRICH, W. W. 1938. The effect of water supply to the tree upon water content, pressure test, and quality of Bartlett pears. Proc. Am. Soc. Hort. Sci. 35, 283–288.

SCHARSCHMIDT, R. K. 1954. Factors affecting the sloughing of canned potatoes. M.S. Thesis, Univ. of Wisc.

SIEGEL, M. 1938. The effect of salt on the drained weight of canned tomatoes. Canner 88, No. 4, 14–15.

SIEGEL, M. 1939. The effects of calcium salts in canning tomatoes. Canner 90, No. 2, 12–13.

SIMON, M., WAGNER, J. R., SILVEIRA, V. G., and HENDEL, C. E. 1953. Influence of piece size on production and quality of dehydrated Irish potatoes. Food Technol. 7, 423–428.

SISTRUNK, W. A., and CAIN, R. F. 1960. Chemical and physical changes in green beans during preparation and canning. Food Technol. 14, 357–362.

SMITHIES, R. H. 1960. Effect of chemical constitution on texture of peas. Soc. Chem. Ind. Monograph 7, 119–127.

SNYDER, E. B. 1936. Some factors affecting the cooking quality of peas and Great Northern type of dry beans. Nebraska Univ. Agr. Expt. Sta. Research Bull. 85.

SON, C. H. 1960. Microscopical study of structural changes of peaches and pears during softening. Dissertation Abstr. 20, 1316.

STERLING, C. 1959. Drained weight behavior in canned fruit: An interpretation of the role of the cell wall. Food Technol. 13, 629–634.

STERLING, C. 1961. Physical state of cellulose during ripening of peach. J. Food Sci. 26, 95–98.

STERLING, C., and CHICHESTER, C. O. 1960. Sugar distribution in plant tissues cooked in syrup. Food Research 25, 157–160.

TORAFSON, W. E., NONNECKE, I. L., and STRACHAN, G. 1956. An evaluation of objective methods for determining the maturity of canning peas. Can. J. Agr. Sci. 36, 247–254.

ULRICH, R., and MIMAULT, J. 1956. Transformation of pectic compounds and respiration of pears during ripening. Fruits 11, 467–470.

VERNER, L. 1931. Experiments with a new type of pressure tester on certain stone fruits. Proc. Am. Soc. Hort. Sci. 27, 57–62.

WECKEL, K. G., SCHARSCHMIDT, R. K., and RIEMAN, G. H. 1959. Sloughing in canned potatoes. Food Technol. 13, 456–459.

WHITTENBERGER, R. T. 1951. Measuring the firmness of cooked apple tissues. Food Technol. 5, 17–20.

WHITTENBERGER, R. T., and HILLS, C. H. 1953. Studies on the processing of red cherries. I. Changes in fresh red cherries caused by bruising, cooling, and soaking. Food Technol. 7, 29–31.

WHITTENBERGER, R. T., and MARSHALL, R. E. 1950. Measuring the firmness of tart red cherries. Food Technol. 4, 311–312.

WILEY, R. C. 1959. Slurry viscosity measurements as methods to determine maturity of lima beans and peas. Food Technol. 13, 694–698.

WOLODKEWITSCH, N. 1956. Methods for the measurement of texture in foodstuffs. Z. Lebensm. Untersuch. u. Forschung. 103, 261–272.

WOODROOF, J. G. 1938. Microscopic studies of frozen fruits and vegetables. Georgia Agr. Expt. Sta. Bull. 201.

CHAPTER 7

Unctuous Foods

INTRODUCTION

The category of unctuous foods includes butter and margarine, chocolate and certain other confections, peanut butter, mayonnaise, ice cream, and similar products. These products are characterized by the presence of a high proportion of fat which acts as the food's major texture-influencing constituent. Fatty foods which are liquid at their usual serving temperature have been discussed in Chapter 3. In keeping with the plan for this volume, raw materials such as the shortenings used in pastry and the like will not be discussed in detail since the primary emphasis is to be on the texture of finished foodstuffs. Some foods contain a high proportion of fat but owe their outstanding textural characteristics to another component. Pie crusts, nuts, and some kinds of confectionery are examples of such products. They will be discussed elsewhere.

BUTTER AND MARGARINE

There are two distinct kinds of texture which affect consumer evaluations of butter and margarine. The first of these is composed of the set of rheological properties determining the ease of cutting and spreading the food when it is at refrigerator temperatures. The second texture is the physical response of the spread as it reaches body heat during ingestion. The former set of qualities has received far more attention than has the latter. Evidently there is good reason to believe that consumers are more responsive to the fluctuations in spreadability which are normally expected in commercial products than they are to "eating texture." It is true, of course, that much of these table spreads are used as a minor cooking ingredient and, in such cases, the mouthfeel has no significance.

The trend of the data in most publications indicates that sectility and spreadability are principally influenced by crystal composition and size as determined by the processing temperatures and other processing conditions. However, rather minor variations in the lipid components of the cream and many other factors have some influence on this kind of texture. Recently, products of improved spreadability have been prepared by whipping gas into the butter to reduce its density.

Butter contains over 80 per cent butterfat, about 16 per cent water, 0.5 per cent protein, 0.5 per cent lactose, and 0.1 to 3.0 per cent ash (mostly from added salt) according to Mulder (1953). The composition of margarine is similar, except, of course, the fat component is not butterfat (some spreads containing small proportions of butter have appeared on the market).

Butter is composed of a continuous phase of butterfat enclosing drops of aqueous solution and globules of lipid material. Crystals or agglomerates of solid fatty substances may also be present in forms sufficiently large to be detected microscopically. Some authorities claim that a second continuous phase composed of aqueous solution exists in butter.

Butterfat is a mixture of various glycerides. The relative proportions of these compounds are controlled by the breed of cattle, the season of the year, and, particularly, the type of feed. The aqueous phase contains milk proteins, native minerals and added salt, lactose, and other skim milk and wash water components. It is the principal source of flavor in butter.

Von Gavél (1956) studied microtome sections of fresh, cooled (32° to 37°F.) butter. He found that the diameters of the air spaces in his samples exceeded those of the water droplets and fat globules. The size of the water droplets was not constant, but varied according to the availability of water. The solution in the droplets seemed to contain hygroscopic substances. Butter having a crumbly structure due to poor working exhibited additional (i.e., non-globular) water in crevices. When studied by phase contrast microscopy in dehydrated sections, the fat globules appeared to be structures embedded in homogeneous fat. Microscopic crystals were observed to exist in the vicinity of large fat globules.

Consistency of butter at any temperature is said by Sumner (1960) to be mostly dependent upon the structure of the fatty phase, i.e., the proportions and distribution of solid and liquid glycerides and the size and nature of the fat crystals. Many investigators have studied the relationship between butter texture and the composition of the butterfat. Coulter and Hill (1934) found that there was a highly significant correlation between the hardness of butter or butterfat and their iodine numbers. Extreme variations in Reichert-Meissl numbers also seemed to be associated with hardness variations.

Mulder (1953) indicated that the direct influence of proteins on the mechanical properties of butter is probably not of great importance and that normal differences in moisture content probably do not affect firmness. However, the degree of dispersion of the water may have some effect on hardness since Mulder has shown that salted butter, which contains relatively large droplets of water, is softer than unsalted butter. Mohr and von Drachenfels (1956A) also discussed the effect of amount and distribution of water on consistency and spreadability. Mohr and von Drachenfels (1956B and C) proposed a scheme for evaluating the water distribution in butter dependent upon measurements of the dimensions of the water droplets. Class I (good) would be comprised of those butters containing no droplets larger than ten microns in diameter, Class II (moderately good) butters could contain several droplets between 10 and 20 microns in diameter, while Class III (poor) products would include numerous droplets larger than ten microns, some being 30 microns or even larger.

Mohr (1957) stated that the firmness of butter depends upon the relative amounts of crystal, gel, and oil in the butterfat and the number of crystal clusters.

The texture of butter has been the subject of a number of intensive investigations. Davis (1937) is said to be one of the first to have applied modern rheological methods to the study of butter consistency. He loaded a cylinder of butter with weights and determined the deformation under these loads and after the loads were removed. He used the permanent deformation measurement to calculate a viscosity by dividing the shearing stress by the rate of deformation. From the recovery measurement Davis calculated a modulus of elasticity for his butter samples. The viscosity and elastic modulus together were considered an indication of firmness.

Kruisheer and den Herder (1938) found that permanent deformation of butter occurs only at pressures exceeding a certain limit. By graphical methods, these workers derived a "yield value" from measurements obtained by use of a simple plastometer. They showed the existence of a clear parallelism between the yield values of several samples of butter and the texture evaluation by experienced graders. Kruisheer and den Herder proposed the use of "yield value" as a quality measurement. They described a mechanically driven device for determining these values.

Dolby (1941A) constructed an apparatus which enabled him to measure the force required to cut a butter sample with a wire. He found that the load required to produce a certain rate of cutting is proportional to the diameter of the wire and the length of the contact between the butter and the wire. According to Dolby (1941B), the rate of shear of butter when cut by a wire increases as a power of the load applied to the wire. Since the values of this power are high, ranging from 8 to 15 (average 11) at 54°F., a high structural viscosity in the butter is indicated. A true yield value for butter could not be detected by this author, but curves described by the rate of shear plotted against the load are such that a practical yield value can be found. Dolby defines the practical yield value as the minimum load required to produce a measurable rate of flow and recommends its use as a convenient measure of the hardness of butter.

The viscosity of butter decreases rapidly as the rate of shear is increased, but the rate of the decrease varies from one sample of butter to another.

Mohr and Wellm (1948) measured the loads needed for penetration of a cone and of a ball into butter. The cone was two cm. high and had a base four sq. cm. in area, while the ball had a radius of two cm. They state that the layer of the butter which is in contact with the penetrometer has reached its yield value when the rate of penetration is zero. With their apparatus, about six hours are required for the rate of penetration to reach zero. However, they suggested an alternate method in which penetration after 60 seconds is recorded provided it does not exceed one cm. Mohr and Wellm derived the following equation for calculating yield value:

$$\text{Yield value} = \frac{\text{Weight}}{\text{Area of contact of cone or ball}}$$

Kapsalis et al. (1960) developed an instrument called the "Consistometer" for measuring both the spreadability and the hardness of butter samples. This instrument was somewhat similar in principle to the one described earlier by Huebner and Thomsen (1957A) which in turn was based upon a device constructed by Mohr and Haesing (1949). The Kapsalis et al. apparatus consists of a constant speed motor driving a pendulum which carries the knife or wire used to measure the spreadability or hardness, respectively. The butter is held in a stainless steel frame mounted at the lowest point of the arc described by the pendulum. The sample is raised one-sixteenth inch for the knife and one-eighth inch for the wire before each determination. The resistance offered by the sample is equivalent to the sum of the forces exerted by fixed weights and the motor. The weights act through a pulley system and can be measured directly, while the force exerted by the motor is registered on a torque-meter. Kapsalis et al. gave an analysis of the reproducibility of the instrument which can be summarized by saying that it appears to be quite accurate enough for most investigative purposes. A diagram of the Consistometer is shown in Fig. 21.

Comparison of Consistometer measurements and consumer panel evaluations of 109 commercial butter samples obtained from 14 states located in different parts of the country revealed a high degree of correlation. The results also indicated that the most desirable butter consistency from the consumer's standpoint is represented by a range on the Consistometer of 400 to 900 gm. resistance to the knife or 140 to 200 gm. resistance to the wire, under the conditions of the study. The instrument differentiated between butters with 120 to 500 gm. hardness and 260 to 1,542 gm. of

spreadability.   These ranges encompass samples which consumer panels
would rate from easy-to-spread to difficult-to-spread.

Using the methods described above, as well as other procedures, the
effects of variations in processing methods and storage conditions on
butter texture have been investigated by many workers.   It appears that
the state of the fat in the butter is determined not only by the temperature at the moment of testing but also to a considerable extent by the
temperature history of the butter before, during, and after processing.

Valentine (1935) found that the rate of cooling of cream had more
effect on the spreadability of butter than did the temperature to which

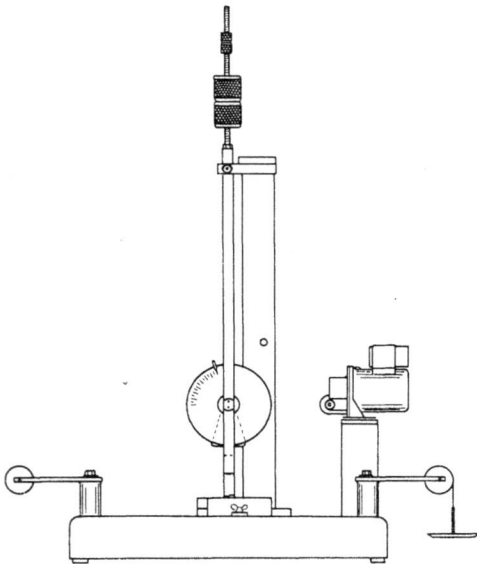

Courtesy Accurate Manufacturing Co.

FIG. 21.   THE CONSISTOMETER—AN
APPARATUS FOR MEASURING THE
SPREADABILITY AND HARDNESS OF
BUTTER

the cream was cooled.   Dolby (1941C) studied the effect on the consistency of butter of the type of pasteurizer, the rate of cooling after
pasteurization, the temperature during holding, the churning temperature,
the temperature of the wash water, and the amount of working.   Only the
rate of cooling after pasteurization had a marked and consistent effect on
the texture of the butter.   Rapidly cooled cream yielded the harder butter.
Lowering the temperature of the wash water caused a small and rather
inconsistent decrease in hardness of the butter.

The effect of changes in the rate of cooling after pasteurization has
been studied by several later workers.   In a statistical study of the 109
samples of commercial butter described by Kapsalis et al. (see Table 11),
it was found that there was a statistical correlation between butter

spreadability and (1) methods of pasteurization, (2) churning temperature, and (3) butter storage temperature. Huebner and Thomsen (1957B) showed that the temperature of the cream after pasteurization and the temperature of butter during the storage and printing procedure had the greatest influence on hardness and spreadability of butter. Storing cream at temperatures above 48°F. after pasteurization and butter at temperatures below 25°F. made an improvement on the consistency of the butter.

TABLE 11

SPREADABILITY AND HARDNESS OF BUTTERS AS DETERMINED BY THE CONSISTOMETER AND A CONSUMER PANEL[1]

Number of Samples	Consistometer Data[2]				Consumer Panel[3]	
	Spreadability		Hardness		Spreadability	
	Range	Average	Range	Average	Range	Average
2	206–403	332	120–144	132	9.5–8.8	9.2
2	574–578	576	138–158	149	9.0–8.8	9.2
4	610–690	648	172–197	183	8.1–6.2	7.4
7	724–789	761	176–230	192	7.4–6.0	6.9
13	831–896	869	196–292	222	8.4–4.7	7.0
22	903–991	943	188–273	226	8.4–4.9	6.4
17	1004–1090	1041	200–285	241	7.2–4.5	6.1
16	1102–1183	1137	218–361	268	6.7–4.0	5.8
10	1216–1286	1250	265–430	332	6.4–3.8	5.2
9	1333–1400	1367	292–451	365	6.1–4.5	5.5
5	1426–1440	1435	366–498	448	5.2–4.7	4.9
2	1514–1542	1514	500–500	500	3.4–3.4	3.4
109[4]	206–1542	1017	120–500	258	9.5–3.4	6.2

[1] From Kapsalis et al. (1960).
[2] Consistometer data expressed in grams.
[3] Consumer panels rated butter on an eleven point scale ranging from 11 = very easy to spread, to 1 = very difficult to spread.
[4] All Samples.

There can be very little doubt that the temperature history of the butter exerts its effect on texture by influencing the average crystal size and the size distribution of the fat crystals.

ICE CREAM

Ice cream is a marginal example of an unctuous food. The butterfat content (usually in the range of 12 to 14 per cent) has an important effect on texture. But size of ice crystals, size of lactose crystals, and amount of air beaten into the product (overrun) are also very influential in establishing the mouthfeel.

According to Tressler and Evers (1957), in Chapteer 19 of *The Freezing Preservation of Foods,* a typical mix formula for 200 gallons of ice cream might be as follows:

	Lbs.
Cream (50 per cent fat)	240
Defatted milk	379
Condensed defatted milk	219
Sweetening	155
Gelatin	3
Egg yolk	4
Total	1000

To the above are added flavors and colors. The ingredients are mixed, pasteurized at 155° to 160°F. for 30 minutes, homogenized at pressures of 1500 to 3000 lbs. per sq. in., cooled, "aged" for 4 to 24 hours (sometimes this step is omitted), and then frozen by a process which includes beating air into the freezing mass.

The most extensive investigations and theoretical treatments of the rheology of ice cream products have dealt with the unfrozen mixes (Sherman 1961; Scott Blair 1953). Although properties of the mix are important in controlling behavior during processing, and can, in certain circumstances, affect the texture of the finished product, the texture of ice cream cannot normally be predicted from an examination of the physical properties of the mix. Variations in mouthfeel are more likely to be established during processing.

Of the ingredients, butterfat, lactose, sucrose, emulsifiers, and stabilizers are the most effective in influencing texture of ice cream. Nickerson and Pangborn (1961) studied the effect of variations in sucrose and milk solids not fat on several properties of a twelve per cent butterfat ice cream. They found that increases in sugar increased the melting rate of the ice cream. The firmness of the product at freezer temperatures also appeared to increase as the sugar content was varied from 13 per cent to 19 per cent. Melting rate and firmness were not much affected by the small changes in amount of milk solids not fat studied by these investigators (see Table 12). The method of Haighton (1959) was used by Nickerson and Pangborn in the determination of the firmness of their samples.

Processing conditions have a marked effect on the textural characteristics of ice cream. The rate of freezing and the amount of overrun are the predominant influences. If the rate of freezing is not sufficiently rapid, large ice crystals may develop. Probably the desired situation is the formation of a crop of very small and uniform (in size) ice crystals at an early stage in the freezing process so that the inevitable growth occurring in the subsequent hardening and storage stages does not lead to grainless.

Graininess can result from the development of large lactose crystals. Removal of the supersaturated lactose in the form of minute crystals during the freezing process reduces the danger of sandiness developing during storage. According to Nickerson (1954), the crystals initially formed should not be larger than ten microns. Large lactose crystals can develop-

TABLE 12

THE EFFECT OF SUGAR AND MILK SOLIDS NOT FAT ON THE VISCOSITY, MELTING RATE, AND FIRMNESS OF ICE CREAM[1]

	13% Sugar, 11% MSNF	15% Sugar, 11% MSNF	17% Sugar, 11% MSNF	19% Sugar, 11% MSNF	15% Sugar, 10% MSNF	15% Sugar, 12% MSNF
Viscosity, in centipoises[2]	136	149	160	230	143	172
Melting rate[3]						
5 min.	0	0	0	0	0	0.5
10	0	2	1.5	2	2	2.5
15	2.5	9	9	9	8	9
20	10	18.5	21	20	20.5	18.5
25	20.5	31	36	32	33	31
30	31.5	41	49	47	45	43
Firmness[4]						
2°F.	12.4	14.4	17.2	17.8	13.1	14.8
12°F.	23.2	27.0	28.0	37.9	26.2	27.4

[1] From Nickerson and Pangborn (1961). The ice cream contained twelve per cent fat and 0.35 per cent 'Hi-gel" stabilizer.
[2] Viscosity measurements made at 40°F. on the unfrozen mix.
[3] Volume in ml. of drip from a cylinder of ice cream 1½ inches high by 3¼ inches in diameter held at 78°F. for the indicated times.
[4] Penetration depth in mm. of a cone-tipped plunger weighing 178 gm.

velop during storage of ice cream because of inadequate number of foci on which the supersaturated lactose can crystallize. Sandiness can be controlled by adjustment of the freezing conditions, or, as suggested by Nickerson, by seeding the mix with lactose crystals.

Nickerson and Pangborn (1961) showed that a straightline relationship existed between per cent overrun and firmness as measured by the penetrometer technique of Haighton. This simple relationship applied only to product made without stabilizer (see Fig. 22).

CHOCOLATE

Chocolate liquor or bitter chocolate is composed of a continuous phase of cocoa-butter enveloping an approximately equal mass of non-fatty particles. There are not present any globules of aqueous solution such as can be found in butter and margarine, although discontinuous fatty regions are detectable. The small quantity of water (less than two per cent) probably exists in association with the particles. Minute air bubbles can be entrapped during processing and will exert an influence on texture. "Eating chocolates" contain fairly large amounts of sugar and/or milk solids in addition to flavoring compounds, emulsifiers (e.g., lecithin) and added cocoa-butter. There is some evidence that sugar or the components of milk solids may, under certain conditions, form a continuous lattice.

The characteristic textural qualities of chocolates are largely a function of the lipid components. Cocoa-butter has a much narrower plastic range than other common natural or manufactured fats. The melting points of most samples of cocoa-butter lie between 90° and 95°F. (Hil-

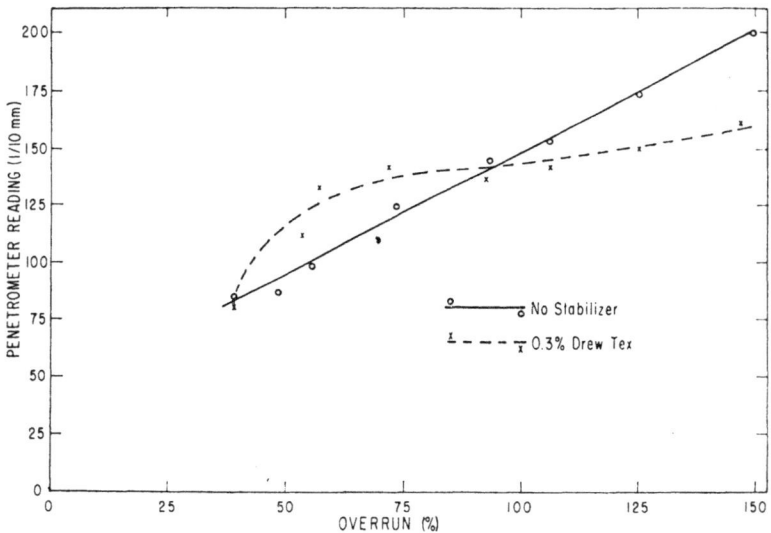

FIG. 22. THE EFFECT OF OVERRUN ON THE FIRMNESS OF ICE CREAM WITH AND WITHOUT STABILIZER

ditch 1956). At room temperature the substance is rather brittle and relatively non-greasy, making for convenient handling. However, it melts at body temperature, readily yielding in the mouth a plastic mass of agreeable texture.

Although the short plastic range of cocoa-butter might seem to suggest that it is composed of glycerides which melt within a narrow range of temperatures, this is not the case. Aylward (1960) published data showing that some (15 per cent) liquid fat is present at 58°F. while some solid material (five per cent) is present at 95°F. Hilditch (1956) gave the following composition for cocoa-butter: palmitostearins, 2 per cent;

oleopalmitostearin, 52 per cent; oleodistearin, 19 per cent; oleodipalmitin, 6 per cent; palmitodiolein, 9 per cent; and stearodiolein, 12 per cent. Apparently, cocoa-butter can be made (by special cooling methods) to assume at least four polymorphic forms which vary in "melting point" from 64° to 94°F. These forms may be due to the existence of liquid micelles in cocoa-butter receiving the special treatments.

The size of the particles of non-fatty substances in chocolate must be reduced to less than about 30 to 60 microns if acceptable texture is to be achieved. Granules with dimensions exceeding about 30 to 60 microns can be detected as "grittiness" in the mouth. It is not sufficient that the average particle size lie within the desired range, instead virtually all of the non-fatty particles must be below the limit in all of their dimensions. Even relatively small numbers of large particles can contribute a gritty feel to foods.

Size reduction in chocolate manufacture is generally accomplished by using steel roll refiners, and one of the purposes of the prolonged conching step is to completely disperse all aggregates of particles which might be perceived as grittiness. Koch (1956) reviewed the methods by which particle size in cholcolate can be measured. He prefers direct measurement by micrometer "when the particle is free to align itself to the jaws of the micrometer." Under these conditions, the shortest side of the largest particle in the mass between the jaws is determined. This is a more useful measurement than average dimensions determined by, say, sedimentation tests. It does require some experience and skill, however. Direct measurement by microscopic examination is sometimes used. As pointed out by Koch, the longest side of any individual particle is measured by this technique.

Although particle size of chocolate is usually in the subsieve range, coarse liquor may be tested by sieving. The usual procedure is to mix the chocolate with an excess of fat solvent, pour the suspension onto a sieve, and wash the residue repeatedly with solvent. A standard 200-mesh screen has openings measuring about 74 microns and thus will retain practically all particles which can be perceived as grittiness. Of course, the sieve measures only the smallest cross section of a particle. Particle length in chocolate can vary from 1 to 8 times the measurement of either of the other two dimensions, according to Koch. The usual ratio of about 4 to 1 is affected by the temperature of the refiner rolls. Berg and Kovac (1960) suggested use of the Coulter counter for determining the particle size in chocolate.

The particle-size-distribution affects the viscosity of molten chocolate and presumably has some influence on eating texture. Increasing the surface area of the non-fatty materials develops greater adsorptive capacity for lipid materials with a consequent reduction in the amount of free or "flowable" cocoa-butter. The net result is an increase in viscosity of the chocolate liquor. The degree of irregularity of the particle surfaces doubtless influences the absorptive capacity, but there seems to be no publication on this subject.

Most of the studies on rheological properties of chocolate have involved measurements made on the melted substance. The most obvious usefulness of such data is in the processing area. For example, the physical behavior of molten chocolate is of particular importance when it is to be used for coating purposes. It would also appear that the characteristics of the molten chocolate are related in some degree to the perceived texture of the finished product. For that reason, the subject will be treated in the present discussion.

Fincke and Heinz (1956) used the Rotavisko viscometer to study the rheology of chocolates. They observed a distinct structural viscosity in all samples examined. Rheopexy and thixotropy were occasionally observed, but only under special experimental conditions. They concluded that melted chocolate represents a complicated rheological system which exhibits many flow anomalies.

The instrument most frequently used in this country for measuring the viscosity of molten chocolate is the MacMichael viscosimeter. Capillary viscometers and devices of the falling sphere type have also been used.

The viscosity of chocolate liquor decreases with increasing temperature. Milk chocolate, and, to a lesser extent, sweet chocolate tend to exhibit an increase in viscosity (after an initial small decrease) when heated. Addition of small amounts (0.3 to 0.4 per cent) of the emulsifier lecithin causes a considerable decrease in viscosity (Schoen 1951) a discovery usually attributed to Rewald of Germany. This effect is most pronounced in systems with low ratios of cocoa-butter to non-fatty substances. Aylward (1960) listed the following as possible explanations of the mechanism of action of lecithin: (a) the lecithin may form a monomolecular film on the surface of the non-fatty particles thereby promoting the wetting of the sugar and cocoa solids by the fat and reducing internal friction; (b) lecithin may displace fat from the surface of the solids and increase the amount of free fat; (c) the emulsifier might minimize the possibility of the adherence together of sugar or other solid particles; (d) lecithin may reduce the size of the particles by a peptization reaction and thereby create a more uniform dispersion; or (e) the emulsifier might "take up" any water present with a resultant dehydration of the cocoa butter and the formation of a hydrated lecithin film on the surface of the non-fatty particles. Hypotheses (d) and (e) seem to be less likely to be correct, the former because the usual effect of a decrease in particle size seems to be an increase rather than a decrease in viscosity, and the latter because the action of lecithin does not seem to be related to the amount of water present in the chocolate.

The work of G. O. Young and L. E. Campbell (reviewed by Harvey 1953) showed that the viscosity of molten chocolate is influenced by a number of processing conditions. The viscosity decreased with increased degree of roasting, partly due to an increased amount of available fat and partly due to some unknown causes. There was an increase in viscosity of milk chocolate upon prolonged conching, but the usual result of an increase in conching time was a decrease in viscosity.

Heiss and Bartusch (1956) listed duration and time of conching, the aging of the mass, and the temperature and degree of seeding as processing conditions which affected the viscosity of chocolate. Of course, temperature and degree of seeding will not affect the viscosity of fully molten chocolate. Sterling and Wuhrmann (1960) showed that purified cocoa butter is characterized by a slight structural viscosity, even in the absence of detectable crystals. As fat crystals form in the cocoa butter, the structural viscosity effect is enhanced. Sub-microscopic aggregation of fat molecules during standing of the cocoa butter is evident by an augmentation of the viscous "coefficient" before microscopically visible fat crystals appear. Some of their data bearing on this point are given in Table 13.

TABLE 13

APPARENT VISCOSITY OF COCOA BUTTER AS A FUNCTION OF TEMPERATURE AND TIME OF STANDING[1]

Temperature, °F.	Average Time of Standing, Hr.					
	1	3	5	7	9	13
86.0	72.2	72.7	74.2	74.9	75.4	75.4
82.4	86.4	88.3	88.8	89.5	89.5	91.0
80.6	98.6	102.0	102.5	103.0	103.4	106.1
78.8	111.4	112.6	113.0	114.7	117.1	121.7
77.0	124.1	126.5	126.5	131.5	143.3	148.6
75.2	138.0	139.9	140.2	153.4	164.6	209.3

[1] From Sterling and Wuhrmann (1960). Viscosity expressed as centipoises.

Both the initial texture and the texture stability are strongly influenced by the final solidification conditions of the finished product. In practice, this means that continual agitation combined with careful control of the rate of heat transfer is necessary. The conditions of "setting" or solidification also control the gloss of the finished chocolate. Unfortunately, the conditions leading to maximum gloss appear to yield the least stable product. A compromise must be sought to yield a chocolate of moderately good gloss with a stability adequate for the storage conditions which the product is expected to encounter.

Improper processing and storage conditions, alone or in combination, can cause the development of the serious texture defect known as "bloom." This texture deterioration process, which is accompanied by the appearance of dull light-colored patches on the surface of the chocolate, is evidently due to the agglomeration or separation of sugar crystals (in "sugar bloom") or of higher-melting forms of some of the triglycerides of cocoa-butter (in "fat bloom"). The ultimate result is a crumbly or gritty textured product which has completely lost the desirable smoothness of good chocolate.

OTHER UNCTUOUS FOODS

Peanut butter is an unctuous food comparable in many respects to chocolate. Both products are prepared by grinding roasted seeds of high fat content. However, the oil which constitutes about 46 to 50 per cent of the mass of peanuts melts at about 46°F. Because of the much lower melting point of its lipid component, peanut butter is a plastic mass at room temperature instead of the brittle solid of chocolate liquor.

Peanut butter is prepared by hulling the nuts, roasting them, removing the testa, and then grinding to a smooth paste. A mixture of Spanish and Virginia peanuts is commonly used, since a butter made of Spanish peanuts is too oily, and one made only of Virginia peanuts is too dry.

If the perception of grittiness during consumption is to be avoided, the non-fatty particles in peanut butter must be ground to a size of less than about 30 to 60 microns. In practice, this degree of fineness is probably achieved only rarely. There is a good demand for "chunky" peanut butter which is prepared by blending a small proportion of relatively large peanut pieces into a mass of nuts ground in the usual manner.

The amount and fluidity of the oil can result in separation at warm temperatures, with resultant unpleasant texture effects. This occurrence can be prevented, or at least retarded, by partial hydrogenation of all or part of the oil and the addition of emulsifiers. Hydrogenation also lessens the tendency of the oil to coat the oral cavity and reduces the unpleasant sensation of the food clinging to the palate during chewing. According to Woerfel (1960), peanut oil contains the following saturated fatty acids in its glycerides: palmitic 7 per cent, stearic 4 per cent, arachidic 3 per cent, behenic 2 per cent, and lignoceric 2 per cent. The unsaturated fatty acids are oleic (60 per cent) and linoleic (22 per cent). The effect of increases in total oil content are to produce a smoother and less sticky product (Morris et al. 1953).

Other nut butters (pecan, etc.) have been made but have attained little popularity.

Mayonnaise is a semi-solid fat-in-water emulsion and could, perhaps, be regarded as a borderline case in the unctuous food category. By Federal Standards, this product must contain not less than 50 per cent oil. Other ingredients are whole eggs or egg yolks (usually the latter), vinegar or lemon juice, and salt. Sugar and spices are sometimes added. The consistency or texture of emulsions of this type is determined principally by the viscosity of the continuous phase, the dimensions of the discontinuous elements (fat globules), and the attractive forces between the particles. Apparently, the viscosity of the fat plays a lesser role. There is a dearth of published reports on the measurement and control of texture in mayonnaise. Sumner (1960) provided a brief discussion and Lowe (1943) reviewed the literature on the subject. The conclusions of Sherman (1961) with respect to the physical properties of emulsifier films at the oil-water interface in ice cream mixes seem to be applicable to mayonnaise as well.

SUMMARY

The texture of butter is determined principally by the glyceride composition and the physical state of the butterfat. The most common methods of objectively measuring butter consistency involve deformation of samples under known forces applied by a "knife" (spreadability) or by a wire (sectility). Such measurements, if properly conducted, can be used with good accuracy to predict judgments of firmness made by consumer panels.

The short plastic range of cocoa-butter contributes to the good handling properties of chocolate at room temperature and the prompt liquefaction of the substance in the mouth. This quality of the lipids of chocolate is not due to a restricted melting range of the fat components but may result instead from some special crystal form assumed by the glycerides. The presence of non-fatty particles exceeding 60 microns in size leads to perception of a gritty sensation when the chocolate is consumed. Texture of the finished chocolate is highly dependent upon the conditions under which solidification occurs. Improper processor storage conditions lead to the development of bloom, a texture defect characterized by a crumbly or gritty structure.

BIBLIOGRAPHY

AYLWARD, F. 1960. Chocolate. Soc. Chem. Ind. Monograph 7, 75–88.
BACKES, J. V., and WILLIAMS, C. T. 1958. Sugar confectionery, chocolate, jams, jellies. *In* Food Industries Manual. Chemical Publishing Co., New York.

BERG, R. H., and KOVAC, G. M. 1960. Here's how to clinch vital particle-size control. Food Eng. *32*, No. 5, 40–43.

BOYLE, J. L. 1959. The stabilization of ice cream and ice lollies. Food Technol. Aust. *11*, 543–551.

BROWN, R. W., HUMBERT, E. S., and GIBSON, D. L. 1959. Ice milk. Development of a formula satisfactory for both soft-serve and hardened product. Can. Dairy Ice Cream J. *38*, No. 5, 66, 68, 70, 72.

COLE, L. J. N., KEUEPFEL, D., and LUSENA, C. V. 1959. Freezing damage to bovine cream indicated by release of enzymes. Can. J. Biochem. Physiol. *37*, 821–827.

COULTER, S. T., and COMBS, W. B. 1936. A study of the body and texture of butter. Minn. Univ. Agr. Expt. Sta. Tech. Bull. *115*.

COULTER, S. T., and HILL, O. J. 1934. The relation between the hardness of butter and butterfat and the iodine number of the butterfat. J. Dairy Sci. *17*, 543–551.

DAVIS, J. G. 1937. The rheology of cheese, butter, and other milk products. J. Dairy Research *8*, 245–264.

DE MAN, J. M., and WOOD, F. W. 1958. Hardness of butter. I. Influence of season and manufacturing method. J. Dairy Sci. *41*, 360–369.

DE MAN, J. M., and WOOD, F. W. 1959. Influence of temperature treatment and season on the dilatometric behavior of butterfat. J. Dairy Research *26*, 17–23.

DOLBY, R. M. 1941A. The rheology of butter. I. Methods of measuring the hardness of butter. J. Dairy Research *12*, 329–336.

DOLBY, R. M. 1941B. The rheology of butter. II. The relation between the rate of shear and shearing stress. J. Dairy Research *12*, 337–343.

DOLBY, R. M. 1941C. The rheology of butter. III. The effect of variation in butter-making conditions on the hardness of butter. J. Dairy Research *12*, 344–350.

DUCK, W N. 1961. Bloom inhibited chocolate and method of producing same. U. S. Patent 2,979,407. April 11.

FINCKE, A. 1956. Rheological studies on chocolates. Zucker-u. Süsswarenwirtsch. *9*, 629–633, 677–679, 725–727.

FINCKE, A., and HEINZ, W. 1956. Studies on the rheology of chocolates. Fette Seifen Anstrichmittel *58*, 902–906.

HAIGHTON, A. J. 1959. The measurement of the hardness of margarine and fats with cone penetrometers. J. Oil Chemists Soc. *36*, 346–355.

HEISS, R., and BARTUSCH, W. 1956. Rheological properties of dark chocolate masses. Fette Seifen Anstrichmittel *58*, 868–875.

HILDITCH, T. P. 1956. Chemical Constitution of Natural Fats. Chapman and Hall, London.

HUEBNER, V. R., and THOMSEN, L. C. 1957A. Spreadability and hardness of butter. I. Development of an instrument for measuring spreadability. J. Dairy Sci. *40*, 834–838.

HUEBNER, V. R., and THOMSEN, L. C. 1957B. Spreadability and hardness of butter. II. Some factors affecting spreadability and hardness. J. Dairy Sci. *40*, 839–845.

KAPSALIS, J. G., BETTSCHER, J. J., KRISTOFFERSEN, T., and GOULD, I. A. 1960. Effect of chemical additives on the spreading quality of butter. I. The consistency of butter as determined by mechanical and consumer panel evaluation methods. J. Dairy Sci. *43*, 1560–1569.

KING, N. 1955A. The physical structure of butter. I. Dairy Inds. *20*, 311–314.

KING, N. 1955B. The physical structure of butter. II. Dairy Inds. *20*, 409–412.

KOCH, J. 1956. Particle size reduction viewed as a unit process. Mfg. Confectioner *36*, 23, 25, 27, 28.

KRUISHEER, C. I., and DEN HERDER, P. C. 1938. Investigations on the consistency of butter. Chem. Weekblad *35*, 719–730.

LAGONI, H., and SAMKAMMER, E. 1956. Some experiments on the relationship between the chemical nature of butterfat and the consistency of butter. Milchwissenschaft *11*, 463–466.

LOSKA, S. J., JR., and JASKA, E. 1957. A disk rheometer applicable to measuring shortening flow properties. J. Am. Oil Chemists Soc. *34*, 495–500.

LOWE, B. 1943. Experimental Cookery from the Chemical and Physical Standpoint. Third Edition. John Wiley and Sons, New York.

LUTTON, E. S. 1957. On the configuration of cocoa butter. J. Am. Oil Chemists Soc. *34*, 521–522.

MOHR, W. 1957. Improvement of the consistency and the structure of butter in the summer and in the winter by the technical churning process. Fette Seifen Anstrichmittel *59*, 217–221.

MOHR, W., and HAESING, J. 1949. Consistency of butter. III. Milchwissenschaft *4*, 255–260.

MOHR, W., and SCHULZ, F. 1948. Consistency of butter. II. Milchwissenschaft *3*, 362–366.

MOHR, W., and VON DRACHENFELS, H. J. 1956A. Consistency of butter and margarine. Fette Seifen Anstrichmittel *58*, 609–613.

MOHR, W., and VON DRACHENFELS, H. J. 1956B. The microscopic determination of the water distribution in butter, a method for the recognition of the correct working and forming procedure of butter. Milchwissenschaft *11*, 126–132.

MOHR, W., and VON DRACHENFELS, H. J. 1956C. Observations on frozen sections of butter as a method for the study of the disposition of crystals and distribution of water in butter. Milchwissenschaft *11*, 228–234.

MOHR, W., and VON DRACHENFELS, H. J. 1957. Consistency and pre-hardening of summer and winter butter, prepared by different processes. Milchwissenschaft *12*, 46–50.

MOHR, W., and WELLM, J. 1948. Viscosity measurement of butter. Milchwissenschaft *3*, 181–185.

MORRIS, N. J., and FREEMAN, A. F. 1954. Peanut butter. VI. The effect of roasting on the palatability of peanut butter. Food Technol. *8*, 377–380.

MORRIS, N. J., WILLICH, R. K., and FREEMAN, A. F. 1953. Peanut butter. III. Effect of roasting, blanching, and sorting on the content of oil and free fatty acids of peanuts. Food Technol. *7*, 366–369.

MULDER, H. 1953. The consistency of butter. *In* Foodstuffs: Their Plasticity, Fluidity, and Consistency. Edited by G. W. Scott Blair. Interscience Publishers, New York.

NAIR, J. H., and MOOK, D. E. 1933. Viscosity studies of fluid cream. J. Dairy Sci. *16*, 1–9.

NICKERSON, T. A., and PANGBORN, R. M. 1961. The influence of sugar in ice cream. III. Effect on physical properties. Food Technol. *15*, 105–106.

PARSONS, C. H. 1940. Crumbly, sticky butter. Natl. Butter Cheese J. *31*, 4–9.

PERKINS, A. E. 1914. An apparatus and method for determining the hardness of butterfat. Ind. Eng. Chem. *6*, 136–141.

REINART, A., and NESBITT, A. 1959. Improved methods for making winter and summer butter. Can. Dairy Ice Cream J. *38*, No. 9, 54, 56, 58, 59.

RIEL, R. R. 1959. Specifications for the spreadability of butter. J. Dairy Sci. *42*, 899–907.

SCHOEN, M. 1951. Confectionery and cacao products. *In* The Chemistry and Technology of Food and Food Products. Edited by M. B. Jacobs. Interscience Publishers, New York.

SCOTT BLAIR, G. W. 1938. The spreading capacity of butter. J. Dairy Research *9*, 208–215.

SCOTT BLAIR, G. W. 1953. Rheology of milk, cream, ice-cream mixes, and similar products. *In* Foodstuffs: Their Plasticity, Fluidity, and Consistency. Edited by G. W. Scott Blair. Interscience Publishers, New York.

SHERMAN, P. 1961. Rheological methods for studying the physical properties of emulsifier films at the oil-water interface in ice cream. Food Technol. *15*, 394–399.

STERLING, C. 1957. Retrogradation in a starch jelly candy. Food Research 22, 184–191.

STERLING, C. 1960. Rheology of cocoa butter. III. Crystalline changes during storage at various temperatures. Food Research 25, 770–776.

STERLING, C., FUMINAGA, S., and WUHRMANN, J. J. 1960. Rheology of cocoa butter. II. Effect of storage temperature on apparent viscosity. Food Research 25, 630–633.

STERLING, C., and WUHRMANN, J. J. 1960. Rheology of cocoa butter. I. Effect of contained fat crystals on flow properties. Food Research 25, 460–463.

SUMNER, C. G. 1960. Emulsions and related dispersions. Soc. Chem. Ind. Monograph 7, 14–19.

THOMSEN, L. C. 1955. Effect of variations in the manufacturing process on body and texture of butter with special emphasis on spreadability. Milk Products J. 46, No. 10, 20–23.

TRESSLER, D. K., and EVERS, C. F. 1957. The Freezing Preservation of Foods. Avi Publishing Co., Westport, Conn.

VALENTINE, G. M. 1935. The spreadability of butter. I. The relation between the rate of cooling. New Zealand J. Sci. Technol. A-16, 206–212.

VON GAVÉL, L. 1956. The structure of butter on the basis of observations on microtome sections. Z. Lebensm. Untersuch. u. Forsch. 104, 1–21.

WECKEL, K. G. 1939. Variability in physical properties of Wisconsin butter. Natl. Butter Cheese J. 30, No. 12, 63–65.

WIESE, H. F., NAIR, J. H., and FLEMING, R. S. 1939. Improvements in the viscosity of pasteurized cream through subsequent heat treatment. J. Dairy Sci. 22, 875–881.

WILSTER, G. H. 1958A. Butter with superior spreading quality obtained with new method. Proc. State Coll. Wash. Inst. Dairying 27, 17–21.

WILSTER, G. H. 1958B. Smooth spreading butter. Milk Prods. J. 49, No. 7, 8–29.

WOERFEL, J. B. 1960. Shortenings. In Bakery Technology and Engineering. Edited by S. A. Matz. Avi Publishing Co., Westport, Conn.

WOERFEL, J. B., and BATES, R. W. 1958. Blending and measurement of the properties of shortening. Food Technol. 12, 674–676.

CHAPTER 8

Friable Foods

INTRODUCTION

According to the method of classification used in this book, friable foods are those foods composed of small, usually irregular, pieces rather loosely bound to one another either with air spaces intervening or imbedded in a common solid or plastic matrix having different textural qualities. Thus they are non-homogeneous on a macroscopic scale and when masticated they readily break down into their constituent pieces. Crisp or flaky products such as cookies, pie crusts, confections such as spun candies, ready-to-eat breakfast cereals, compressed cereal survival bars, and many representatives of the group called snack foods are common examples of the category.

The essential textural features of friable foods are that little or no sensation of elasticity or resilient yielding is present, and that the product, being porous, or with other large scale discontinuities, readily breaks down into numerous irregular particles during mastication. In common terminology, these products are "crunchy."

STRUCTURE IN FRIABLE FOODS

Cookies of the sugar-snap variety provide good illustrations of the type of structure found in the foods of this class. The dough from which the baked product is made is non-uniform, possessing discrete globules of fat which are often of visible size. Upon baking, the leavening gases (water vapor and carbon dioxide) fill the interior of the dough mass with bubbles, which are probably formed initially at lipid-aqueous interfaces. There are no protein-lined spheroidal vacuoles formed as there are in breads and the like. As the temperature continues to rise in the dough piece, the starch partially gelatinizes, the protein coagulates, and the expanding bubbles of gas form strata or irregular vacuoles partially separating pieces of dough. As the dehydration progresses into its final phases, the dough pieces contract, but, being plastic because of their high temperature, do not separate, and the gas spaces become continuous with the atmosphere. When the product has cooled, the small dough masses remain adherent at their contiguous boundaries and remain separated in the regions where gas pockets had formed. The dough itself becomes brittle and hard. The ease with which the cookie breaks into pieces when pressure is applied to it is a function of the size and

143

shape of the gas pockets, the moisture content and other crumb characteristics, and the strains set up during cooling.

Not all cookies can be classified as friable foods. Some of these products are chewy. That is, their crumb possesses a definite sensible elasticity which prevents a rapid breakdown into small particles when stress, such as chewing, is applied. The composition of the product, especially its moisture content and the type and amount of sugars and dextrins, determines whether it will be friable or chewy. Generally speaking, moisture contents of five per cent or less lead to the friable texture which is the most common in cookies, while moisture contents of ten per cent or more tend to cause a chewy or elastic effect particularly if present in conjunction with large amounts of dextrins.

PERCEPTION AND MEASUREMENT OF TEXTURE IN FRIABLE FOODS

The perceived texture of a friable material will depend upon the forces holding the particles together and the size, shape, consistency, and degree of uniformity of the particles resulting when the piece is broken down. Factors in the consistency of the individual particles which affect the over-all texture include the smoothness or roughness of their surfaces and their hardness.

MEASUREMENTS OF THE TEXTURE OF FRIABLE FOODS

Measurements of the texture of friable foods have generally been made by methods which determine the force necessary to break a piece of the material. The piece size and boundary characteristics of the broken segments have generally not been taken into consideration although these factors undoubtedly affect perceived texture. It is true that some investigators have used destructive sieving tests on, for example, corn flakes, in order to determine the rate of breakdown in piece size, but these studies were usually related only to the resistance of the product to handling and shipping stresses.

An approach, which seems not to have been taken but which would clearly provide much useful information related to sensory evaluations, would be to measure the particle size of the debris resulting from the application of crushing pressures to moderate volumes of the food material. For example, the application of force at a constant rate to a quantity of cookies, cornflakes, potato chips, or popcorn contained in a cylinder with a subsequent particle size analysis of the debris should give a quantitative indication of eating quality. Charting of the course of the resistance to crushing force would give still more valuable information. Of course, the contacting surfaces and the force vectors in the test chamber would vary greatly from place to place in cylinder and from time to time during the test, but the over-all measurement should be fairly precise if a sufficiently large sample is taken.

Davis (1921) described a device called a "shortometer" which he had designed for measuring the breaking strength or crushing strength of baked products such as cookies. Two horizontal parallel rails supported the sample. An upper bar parallel to the lower bars and midway between them applied the crushing force. The force was varied by adding shot to a container, or, in another model, by sliding a weight along a bar. After conditions had been standardized, 30 replicate determinations provided a satisfactory precision. The lower the load required for breaking, the "shorter" the cookies. The "shortening values" of many different cooking fats were determined by this method and Davis found lard to be the most effective of these.

Bailey (1934) improved the shortometer of Davis and added some original features. In his version, the cookie was laid on two supporting rails bolted to the platform of a spring scale. The rails were made of round iron bars 6.5 mm. in diameter which were mounted 40 mm. apart. A similar rail or bar mounted above and midway between the lower supports was the striking member. The upper rail was mounted on a lever moved at a constant rate by a string drawn by a motor geared down to 4.5 rpm. The average probable error of the means of ten series of tests that were each replicated 20 times was 36.2 gm., when the average was about 1,500 gm. Thus the average probable error of the mean of the means was about 2.41 per cent of average.

Binnington et al. (1939) developed a breaking-strength apparatus patterned after the Bailey shortometer. Although they used the improved apparatus for testing alimentary pastes (which will not be considered in this chapter), it should also be suitable for cookie evaluations.

Coppock and Cornford (1960) mentioned a method for measuring the strength of cookies which has found limited use. In this procedure, a pivoted weight is allowed to impinge upon the piece with a hammer-like blow. The cookie is supported on a pair of knife-edges connected with a means for recording on a scale the absorption of energy before the cookie cracks.

STRUCTURE OF FRIABLE FOODS

Wide deviations between individual determinations of breaking strength are often found as a result of "checking" in cookies. Checking is the presence of very fine (sometimes invisible) fractures extending part of the way through the product. It is the result of internal tensions resulting from uneven contraction of the cookie during the final stages of baking. It is most common in hard sweet goods (e.g., English tea biscuits) in which an appreciable development of gluten has occurred. Soda crackers also can exhibit the fault contrary to some statements in the literature.

Unless checking is very extensive, it would not be expected to contribute greatly to organoleptic texture, but it can have a pronounced influence on individual determinations of breaking strength. If several samples from the same batch are available for testing, the influence of checking can be minimized by discarding all samples which have any visual indication of cracks and by eliminating all test values which fall far below the mean. Samples which fracture in a direction or pattern not consistent with the applied forces should not be included in the average.

SOME SPECIAL TYPES OF FRIABLE FOODS

As discussed in the preceding section, one group of friable foods includes those products composed of particles partly bound together and partly separated by air spaces. A second type includes foods composed of solid particles of appreciable size embedded in a matrix having textural qualities perceptibly different from those of the particles. The initial observation of texture in foods of the second type is principally a function of the matrix characteristics. The particles may affect the sensation at this time to some extent, but their major contribution takes effect during latter stages of mastication.

CEREAL BARS

An example of the second type of structure is the compressed cereal bars used by the Armed Forces for special "survival" situations and also useful as concentrated rations for hunters, campers, explorers, etc. They combine high caloric density with extreme storage stability. They consist of a ready-to-eat cereal, such as cornflakes, combined with sugar, fat, and milk solids and compressed into a rectangular bar of (usually) one to two ounces in weight. Instead of being compressed, they may be a very dense baked product, i.e., one in which no leavening action is permitted (Cryns 1956).

In formulating and engineering compressed cereal bars, it is necessary to achieve a bar sufficiently rigid to withstand normal handling abuse during packaging and distribution, but fragile enough to break down readily in the mouth. Experience has shown that a bar which is difficult to bite and chew will often be discarded by the intended consumer even when conditions of sub-normal caloric intake exist. Furthermore, the particles resulting from the initial breakdown in the mouth must be of considerable size. If the bar rapidly degenerates into a powdery or gritty substance when chewed, it will have a very low acceptability. Bars containing gun-puffed cereals behave in this manner and for this reason gun-puffed cereals are not suitable for use as raw materials in compressed cereal bars. To achieve the desired crunchiness, particle fracture during mixing of the ingredients and the compression stage must be minimized. Small flakes and granules have been found to be the most satisfactory forms of cereal.

Matz (1958) described a method of preparing compressed cereal bars by which a finished product held together by a sugar lattice, instead of by a plastic shortening phase, is formed. These bars are generally of superior textural qualities, having a crispness not often achieved in the fat-bound types. Matz also describes a simple apparatus for testing fragility of cereal bars.

Compression does not seem to be as well-appreciated a food processing method as it ought to be. Although cereal bars (and similar foods prepared from potato chips, etc.) are only used in highly specialized situations at present, they could have much wider application. The texture and flavor of cereal bars is, in the best cases, excellent, and the satiety factor is favorable. Similar items formulated as special dietary foods should receive good demand and could be made cheaper and distributed easier than liquid or canned high-moisture dietary supplements. The structural stability of these bars is good so that packaging can be simple and cheap. It is probable that many confections now formed by a combination of melting, mixing, molding, and cooling techniques could be made just as well by mixing and compression with no sacrifice of desirable textural characteristics.

POPPED CORN

Popped corn represents a third type of friable food which also includes the puffed ready-to-eat breakfast cereals such as gun-puffed wheat and rice kernels and corn pieces and oven-puffed rice. The structure of potato chips is somewhat similar. All of these products consist of dehydrated networks of (mostly) protein and starch surrounding a very large volume of air cells.

Popped corn consists of puffed endosperm associated with pericarp pieces of normal density. The puffing results from sudden expansion of steam in vacuoles surrounded by an endosperm plasticized by heat. Simultaneous with, or immediately after, the puffing phenomenon very rapid dehydration of the endosperm occurs, "setting" the kernel in the expanded form. Considering the low density of the puffed material, it has considerable mechanical strength and resists fragmentation very well. From another point of view, popcorn could be considered a food sponge, and some of these aspects of its texture will be discussed in Chapter 10.

In all of the puffed foods retention of crispness is definitely dependent upon the maintaining of low moisture contents. As the "free" water content rises over about 3 to 5 per cent a pronounced decrease in crispness becomes noticeable with an accompanying increase in toughness or elasticity. Since most of these products are quite hygroscopic because of the large surface area and dehydrated condition, they must either be kept in packages made from films resistant to moisture-vapor transmission (and well sealed) or be kept warm. The effect of the latter treatment is, of course, to surround them with air of low relative humidity. An alternate approach is to reheat the product before serving in order to drive off the adsorbed water vapor.

The chief basis for judging the commercial desirability of popcorn has always been the relative amount of kernel expansion that is obtained when the corn is popped. The trade measures expansion by popping a given weight or volume of corn under controlled conditions, dumping the popped kernels into a graduated plastic cylinder of standard diameter, and reading the height of the column. Results are expressed as cubic inches of popped corn obtained per pound of raw corn. Most commercial corn expands between 30- and 35-fold, although some recently developed hybrids give expansion ratios as high as 40-fold. The density of the puffed kernel is clearly related to the texture since the amount of structural material per cubic centimeter will control the resistance to shear, assuming that the structural material and its form are approximately the same from sample to sample.

Pericarp thickness has been suggested as another indicator of popcorn quality (Richardson 1957 and 1958). The pericarp or hull is a thin but tough protective layer surrounding the grain except at the tip. During popping, this tissue is fragmented but most of it adheres to the puffed endosperm. It provides an unpleasant contrast in textures with the crisp endosperm. The thickness of pericarp sections can be measured microscopically. The harsh and unpalatable tip cap also detracts from the texture quality of popped corn. Richardson (1960) indicates that the germ contributes significantly to the texture of popped corn. In the absence of germ tissue the expanded kernel assumes a more elastic quality.

PIE CRUSTS

Pie crusts are low in moisture and high in fat and are made without leavening agents. These formulation characteristics, together with the method of preparation, prevent the formation of a continuous gluten network through the dough mass and result in baked products that are friable or flaky. A porous structure such as typifies all leavened bakery products is not desired because pie crusts must support and retain without leakage fillings of moderate viscosity and high moisture content. Since there are ideally no voids in pie crusts, the structure of these products is importantly different from that of cookies in which the air spaces are significant contributors to texture.

The texture of pie crust is dependent upon the type of flour and shortening and the relative proportions of these two ingredients, the amount of water, and, to a very great extent, the method of preparation. Pie crusts are generally divided into three classes based upon their relative "flakiness." The latter term can be defined loosely as the tendency of the crust to separate into strata or layers when it is broken. Flaky or long-flake crusts tend to exhibit fractures along different lines at different levels when they are broken and to show separation in layers parallel to the surface. Mealy crusts can be broken in a straight line and show a fracture surface more like that of a cookie. Short-flake or flaky-mealy crusts exhibit characteristics intermediate between the two extremes.

The flaky crust is esteemed by connoisseurs and by trade experts, but it shows the abuse of handling more quickly than the mealy type and it poses difficulties in serving. Mealy crusts can be cut readily, and have a texture that is completely acceptable to the average consumer. Most manufacturers catering to the general public make pies with mealy-type crusts. Home recipes are generally designed to produce flaky crusts, but most packaged mixes yield the mealy variety.

The weakest flour available is used for pie crusts because a tough gluten structure must be avoided. In practice, unbleached flours having a protein content of 7.0 to 8.5 per cent and an ash content of 0.38 to 0.48 per cent are often used. The flour should be milled from a soft white winter wheat and should be slightly granular. Soft red winter wheats yield somewhat stronger flours which may or may not be satisfactory for the purpose, depending upon how critical the finished product requirements are.

Shortening for pie crust doughs is used at a level of about 70 to 75 per cent (flour weight basis) in household recipes, about 50 to 60 per cent in commercial pie crusts, 50 per cent or less in doughs made with liquid shortenings, and as little as 35 per cent in fried pie doughs. Lard is generally regarded as the shortening of choice for pie crusts, but satisfactory texture can be obtained with most other types as well.

Flaky crusts are obtained by mixing all of the shortening with all of the flour for a period just sufficient to reduce the shortening to small lumps. The recommended size for the fat pieces varies somewhat according to the authority consulted and the type of shortening employed.

Generally, all of the ingredients are refrigerated, and, in any case the flour and water are brought to 35° to 40°F. before mixing. The doughs are invariably refrigerated before being rolled and cut in order to keep the shortening particles separate and to prevent adsorption of the fat by the flour. The net effect of these techniques is to give flat particles of dough (flour and water) separated by flattened pieces of shortening. On baking, the shortening melts and is ultimately absorbed by the dough but the absence of gluten networks in the layers where the fat formerly was situated causes the development of flakiness.

In preparing doughs for mealy crusts, the shortening may be completely dispersed in the dough. Consequently, processing conditions are less critical. The result of baking this type of dough is a crust which breaks into small irregular pieces. The points of fracture are at discontinuities resulting from the presence in the raw dough of small irregular pieces of shortening.

Texture of pie crusts deteriorates when stored in contact with fillings. Due to its high degree of dehydration, the crust always has a considerably lower equilibrium relative humidity than the filling, so it takes up water and becomes soggy after a short storage period.

There are evidently no publications describing objective texture-testing methods for use with pie crusts although it is generally agreed that the texture has a great effect on the acceptability, not only of the crust, but of the pie. Evaluations have been solely organoleptic in nature. Objective methods of texture determination based upon those suggested for cookies would seem to be possible.

PUFF PASTRY

Puff pastry is a product having a composition similar to that of pie crust but prepared in such a manner that leavening by water vapor creates relatively enormous voids in the finished product. The essential difference in processing, as compared with pie crust dough, is that continuous layers of fat are formed between layers of fairly well developed ("extensible") dough. Schmied (1959) describes six methods by which this may be accomplished. Upon baking steam accumulates between the dough layers and causes them to expand. The finished product fragments readily because of the thinness of the layers and the high shortening content. Texture, then, is determined by ingredient types and proportions, and the thickness of the layers. In turnovers, one of the bakery products having a puff pastry base, consolidation of layers will be observed around the filling. The tenderness of this region has a major effect on acceptability. Thickness of the original dough piece, quality of the dough, amount of filling, and the baking conditions govern the texture of the consolidated layers. As with pie crusts, evaluation of the texture of puff pastries has been entirely subjective in the past.

SUMMARY

Friable foods can be considered to be composed of masses of granules or flakes of appreciable size which are adherent to their neighbors at several points but which are otherwise separated by air-filled spaces or by a matrix having different textural qualities. The granules break apart readily when the food is bitten and chewed, yielding a multitude of crisp particles. Cookies, puffed cereals, and pie crusts are common examples of this type of food.

Textural quality of friable foods is determined by the ease with which the particles separate when masticated, the size and consistency of the resultant granules, and the particle size distribution. Crispness, an essential characteristic for good acceptability, is a result of the details of microstructure and the moisture content, which must be very low.

Measurements of texture in friable foods generally depend upon determinations of the force required to cause breaking under standardized conditions. Microscopic fractures created during manufacture can cause wide variations in the results obtained with duplicate samples, and, for this reason, it is desirable to take the mean of many determinations before making a quality judgment on the basis of the results of breaking tests. Some products such as puff pastry and pie crusts are not tested by objective methods according to current practice.

BIBLIOGRAPHY

BAILEY, C. H. 1934. An automatic shortometer. Cereal Chem. *11*, 160–163.

BINNINGTON, D. S., JOHANNSON, H., and GEDDES, W. F. 1939. Quantitative methods for evaluating the quality of macaroni products. Cereal Chem. *16*, 149–167.

COPPOCK, J. B. M., and CORNFORD, S. J. 1960. Texture in bread and flour confectionery. Soc. Chem. Ind. Monograph 7, 64–74.

CRYNS, J. 1956. Method of making compressed cereal bars for emergency rations and the resulting product. U. S. Pat. 2,738,277. March 13.

DAVIS, C. E. 1921. Shortening: Its definition and measurement. Ind. Eng. Chem. *13*, 797–799.

MATZ, S. A. 1958. Process for manufacturing compressed cereal bars. U. S. Pat. 2,824,806. Feb. 25.

NELSON, O. E., JR. 1955. Purdue hybrid performance tests for 1955. The Popcorn Merchandiser *10*, No. 3, 3–9.

REEVE, R. M., and NEEL, E. M. 1960. Microscopic structure of potato chips. Am. Potato J. *37*, 45–52.

RICHARDSON, D. L. 1957. Purdue hybrid performance trials encouraging. The Popcorn and Concessions Merchandiser *12*, No. 4, 10–17.

RICHARDSON, D. L. 1958. Two factors of early harvesting contribute to popcorn quality. Concessionaire Merchandiser *13*, No. 4, 5, 12–13.

RICHARDSON, D. L. 1960. Pericarp thickness in popcorn. Agron. J. *52*, 77–80.

SCHMIED, K.-H. 1959. Fat distribution in puff paste prepared by different methods. Brot u. Gebäck *13*, 225–233.

YAMAZAKI, W. T. 1959A. Flour granularity and cookie quality. I. Effect of wheat variety on sieve fraction properties. Cereal Chem. *36*, 42–51.

YAMAZAKI, W. T. 1959B. Flour granularity and cookie quality. II. Effects of changes in granularity in cookie characteristics. Cereal Chem. *36*, 52–59.

YAMAZAKI, W. T. 1959C. The application of heat in the testing of flours for cookie quality. Cereal Chem. *36*, 59–69.

CHAPTER 9

Glassy Structured Foods

INTRODUCTION

Hard candies ("boiled sweets") of the clear, transparent type are the only common comestibles which can be included in the category of glassy-structured foods. They are amorphous products, usually translucent or transparent, and are composed of solidified supersaturated sugar syrups. They are glasses in the broadest sense of that word and possess little elasticity, tending to fracture rather than deform when they are subjected to excessive stress.

The distinguishing feature of confections of the glassy type is a continuous homogeneous structure consisting of a nearly dehydrated mixture of low molecular weight carbohydrates devoid of crystalline forms. Minute amounts of coloring and flavoring agents are usually present but have little effect on the texture.

Hard candies might be divided into two classes on the basis of their size: (1) pieces small enough to be taken into the mouth, and (2) pieces designed to be "licked," e.g., lollipops. Textural perception pathways should be similar for products in the two size groups.

It is generally not expected that these foods will be bitten or chewed by the consumer. During consumption, the entire candy piece is slowly dissolved by saliva, continually bringing new surfaces into contact with the mouth and tongue. Texture in these foods is strictly a function of the characteristics of the exposed surfaces except when the structure becomes so degenerate as to allow the piece to crumble.

The initial surface is usually intended to be very smooth, the prevention of any roughness or grittiness being a primary goal of the manufacturing process. In some cases, a decorative pattern may be formed on the surface but the elements of the pattern are made sufficiently large and smooth that they do not create a sensation of grittiness or granularity. It is considered desirable that the structure be uniform throughout the mass so that all surfaces which come into contact with the mouth during the gradual dissolution of the food will be just as smooth as the initial surface. Unlike friable foods, they do not contain textural discontinuities.

For those pieces of candy designed to be taken into the mouth while they are whole, the initial size will play a role in texture perception because the relationship of the size of the piece to the size of the buccal cavity will affect not only the muscle senses but also the force with which the candy is pressed against the tissues of the mouth, greater pressures accentuating the effects of roughness. The shape of the piece must also affect texture. The spherical forms would seem to be less desirable than the flattened forms such as discs and rounded squares of moderate thickness or oblongs of slightly greater thickness. Sharp edges are to be avoided, of course.

EFFECTS OF COMPOSITION ON TEXTURE

Hard candies are made by evaporating concentrated syrups of sucrose mixed with invert sugar and corn syrup. The invert sugar may either be added as such or be formed during processing as a result of hydrolysis of the sucrose. The extent of the hydrolysis may be controlled to some extent by adding acidic buffers such as potassium acid tartrate (cream of tartar). Some hydrolysis occurs even in the absence of added acid components. Colors and flavors are most commonly added after the cooking step is completed in order to minimize thermal destruction and volatilization.

Less than two per cent of water remains in the cooked mass at the time it is formed into pieces. In practice, the moisture content is controlled by adjusting the temperature to which the syrup is brought during the final stages of cooking. Pure anhydrous sucrose melts at 367°F. and begins to caramelize at about 400°F. Due to the extremely viscous nature of the syrups dealt with in this technology, convection currents cannot be relied upon to assure uniform temperature distribution during heating, and most mixing techniques are ineffectual. As a result, some portions of the syrup may be overheated during cooking with the production of sugar decomposition products. This is deleterious to texture as well as to color and flavor. Improvements in cooker design have been directed toward eliminating localized overheating. Removal of water can be facilitated by cooking in vacuum pans, the lowered pressure permitting production of syrups having the desired water content at lower temperatures, thereby minimizing caramelization and other undesirable changes. A recent development is the use of flash evaporation from continuous thin films flowing over the cooking surface. The very short periods at high temperature result in greater flexibility and closer control of processes (Cramer 1961).

The moisture content of the candy has a definite effect on its texture. The presence of four per cent water results in a product which is noticeably softer in the mouth even when the product is not chewed. Hardness increases rapidly as the moisture is brought down from four per cent. At about 1.5 per cent, a peak in hardness is reached which is not much affected by further removal of water.

The crystallization of sucrose from the supersaturated syrup during cooling is inhibited by the "doctors," corn syrup or invert sugar. Corn syrups are clear, colorless, viscous liquids consisting of mixtures of glucose ("dextrose" to the candy maker), maltose, and higher saccharides formed by acid (or combined acid and enzyme) hydrolysis of corn starch. They are sometimes treated with ion exchange resins to reduce the content of inorganic constituents.

Corn syrups may be classified on the basis of their dextrose equivalent, which is a common method for expressing the extent to which the higher saccharides have been hydrolyzed. Dextrose equivalent is defined as the total reducing substances expressed as per cent glucose, on the dry basis. Low conversion syrups have dextrose equivalents (D. E.) of 28 to 38; regular conversion syrups range from 38 to 48 D. E.; intermediate conversion indicates a D. E. of 48 to 58; and high conversion syrups fall within the range of 58 to 68 D. E. All of these types have total solids contents of about 79 to 82 per cent. Dried forms are also available.

Completely hydrolyzed corn sugar (crystalline "dextrose") is available in dried form (about nine per cent moisture) but not as syrups because of the limited solubility of the pure material. An anhydrous form is also manufactured.

The relative proportions of glucose, maltose, and higher saccharides in the corn syrup influence the texture of the candy. Higher levels of dextrins, as found in lower conversion syrups, tend to make the product tougher. Higher levels of glucose and maltose, particularly the former, cause the candy to be more hygroscopic, leading to a greater tendency for it to develop surface stickiness.

Invert sugar (equal quantities of fructose and glucose resulting from the hydrolysis of sucrose) is present in all hard candies since it is practically impossible to heat sucrose syrups to the temperature range necessary for the manufacture of these products without breaking down some of the sucrose. The invert sugar content can be increased either by adding prepared syrups which contain a high concentration of glucose and fructose, or by adding an inverting agent before cooking. The invert sugar content of hard candies usually falls in the range of 3 to 20 per cent. The higher amounts make the product excessively hygroscopic leading to stickiness, while lower amounts may be insufficient to prevent graining (crystallization).

Commercial invert sugar syrups vary in their composition from a small percentage of invert (the rest being unconverted sucrose) to almost complete inversion. They can be made by either an acid or an enzyme process. The latter type is preferable because undesirable thermal changes are minimized. A common commercial type of invert syrup contains about 50 to 60 per cent glucose and fructose on a solids basis and has a total solids content of about 77 per cent. Syrups with a higher proportion of

invert may precipitate glucose. The pH of invert sugar syrups is lower
(about 4.2 to 5.6) than that of sucrose syrups (near 7.0). The ash content may also be slightly higher.

Invert sugar and corn syrup both function efficiently in preventing the
crystallization of sucrose from the cooling mass, but they differ somewhat
in the textural effects they cause in the finished candy. An undesirable
toughness and plasticizing effect (Cramer 1950) can result from the use
of corn syrup and these effects are accentuated as the dextrin content of
the candy is increased. Use of invert sugar alone can cause too much
brittleness and fragility. Both invert sugar and corn syrup increase the
hygroscopicity, and therefore the tendency to develop surface stickiness,
but the former raw material has the more pronounced effect.

According to Mallows (1960), the smoothness of hard candy, (which
he regards as probably the most important aspect of texture in these
foods) is greatly influenced by the method of cooking the syrup. Different
types of cooking apparatus lead to different degrees of localized overheating with varying effects on inversion, caramelization, and other
changes which influence smoothness of the finished candy.

Hard candies are formed into the desired shapes and sizes by either (1)
depositing the hot syrup into molds, (2) running the candy out on to an
oiled slab and cutting the cooling material into the form required, or (3)
using automatic devices which pull the partially cooled but still plastic
mass out in a continuous "rope" from which pieces are cut. Each of these
methods produces a somewhat different texture in the finished product.
The oil from the slab tends to reduce the rate at which stickiness forms
on the surface to which it adheres. Different amounts and sizes of air
bubbles are incorporated into the candy by the different processing
methods.

If air bubbles are present in the mass, they will be perceived as sharp
particles when they become exposed by solution of the surrounding material. The consumer will frequently assume that this sharpness is due to
the presence of foreign particles, i.e., glass. In any event, bubbles contribute an unpleasant textural characteristic to the candy and are to be
avoided.

MEASUREMENT OF TEXTURE

Very few attempts have been made to measure objectively the textural
qualities of hard candies. Harvey (1953) could not find any published
records of rheological measurements on boiled sweets. Perhaps instruments designed for determining the hardness of mineralogical specimens
could be adapted to the measurement of the relative hardness of hard
candy, but this does not seem to have been done. Katz *et al.* (1961)

used a device called a Barcol Impressor to measure the hardness of a foodstuff (endosperm of wheat). The characteristics of this instrument suggest that it might be adaptable to the measurement of the texture of hard candy.

Heiss (1959) described an apparatus for measuring the important textural quality of surface stickiness. This instrument was intended to quantitate the evaluation obtained by pressing a "dry and clean finger" on to the candy momentarily and then pulling it away. In the Heiss apparatus, shown in Fig. 23, an upper plate cushioned with rubber is connected to a calibrated spring. A lower movable plate carries the candy piece which is made perfectly flat on the upper surface. The upper plate

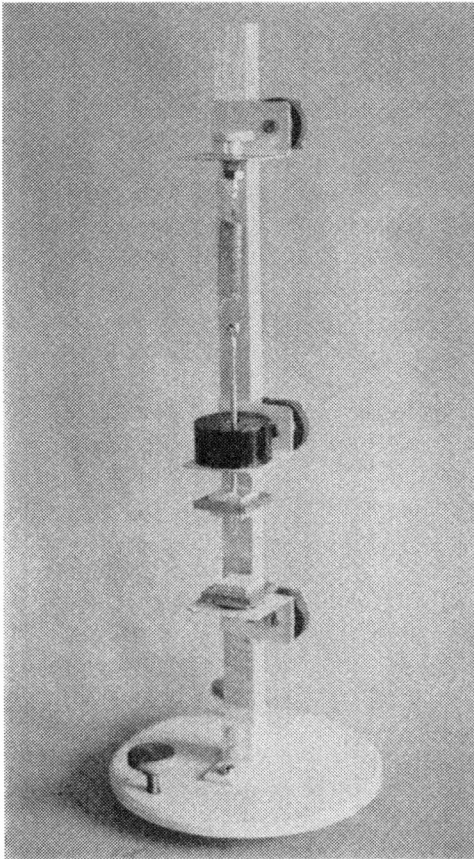

Courtesy Dr. R. Heiss

FIG. 23.   DEVICE FOR MEASURING THE SURFACE
STICKINESS OF CANDY

is initially pressed on to the candy with a weight of 11.5 ounces for ten seconds and then the lower plate is gradually withdrawn at a constant rate of speed until the candy surface and the rubber surface are pulled apart. The distance travelled before the release occurs is read off of a scale inscribed on the upright beam which carries the lower plate. This figure is a measure of the degree of stickiness of the candy surface. According to the results published by Heiss, the instrument gives reliable results.

The thickness of the grained surface of hard candy pieces, which is an indication of texture change during storage deterioration, can be measured microscopically. Fragility, toughness, amounts of large discontinuities such as bubbles, and surface roughness are important textural characteristics for which there are no published objective methods.

STORAGE DETERIORATION

Hard candies are metastable systems which can maintain their characteristics only within a rather limited range of conditions. Because of their highly supersaturated status, they have a pronounced tendency to revert to the more stable crystalline form when conditions which allow the sugar molecules to move into a more ordered array come into being. This movement is restricted in the desired glassy form by the presence of intervening molecules of foreign species and by the limited availability of water molecules. The latter situation not only inhibits the mobility of the carbohydrate molecules but also limits the number of preferred hydrate crystal lattices which can form.

Although increasing the amounts of corn syrup carbohydrates and invert sugar solids tends to prevent crystallization by interfering with the formation of sucrose lattices, these ingredients also tend to make the product more hygroscopic so that, in atmospheres of equal relative humidity, larger quantities of water will be present to mobilize the sucrose molecules. The secret of successful manufacture lies in the ability to choose judiciously the concentrations of these auxiliary ingredients which will yield the highest resistance to graining with the minimum tendency to absorb water. The direct texture effects of invert sugar and corn syrup are relatively minor when compared with the importance of their function as graining inhibitors.

The equilibrium relative humidities of hard candies varies with their composition. Figures of 25 to 40 per cent R. H. are frequently quoted as representative averages. The Heiss (1959) paper previously quoted contains extensive data on the response of hard candy of different composition to variations in the ambient relative humidity (see Table 14 and Fig. 24) and the following discussion is based largely on his material.

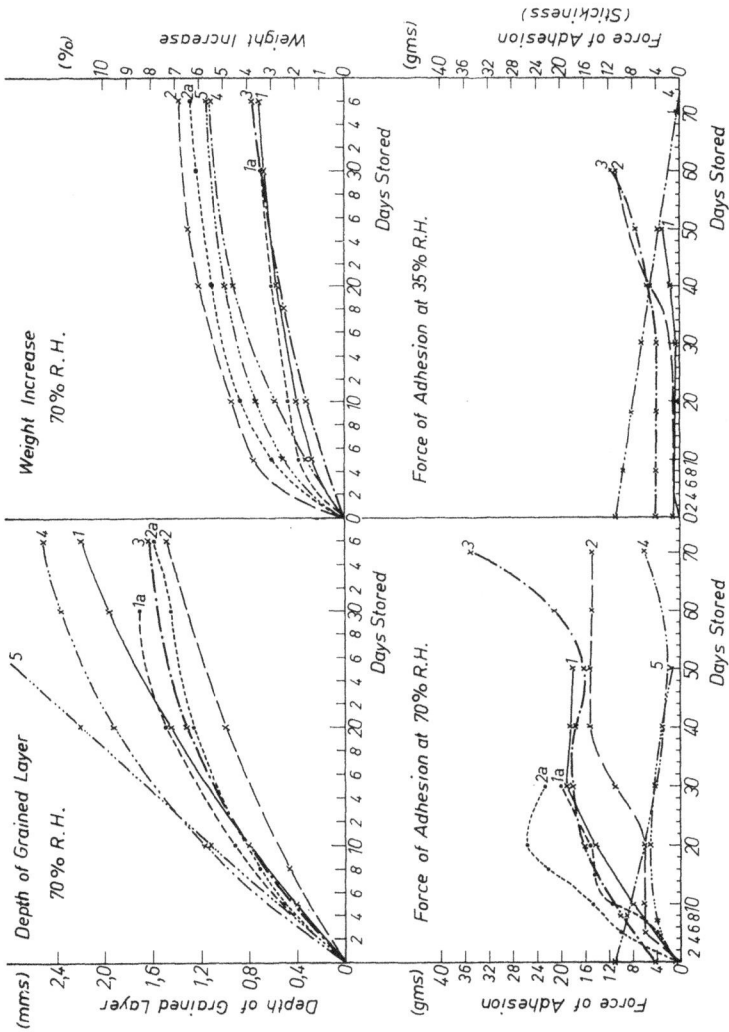

FIG. 24. RESPONSE OF HARD CANDY TO STORAGE AT DIFFERENT RELATIVE HUMIDITIES

Courtesy Dr. R. Heiss

When hard candies are exposed to an atmosphere of higher relative humidity than their equilibrium value at the effective temperature, they will, of course, adsorb water. The initial texture change observed is the development of a sticky surface as a result of the solution of the sugars in the layers near the surface of the candy. This stickiness will continue to increase, if the atmospheric relative humidity is sufficiently high, until enough water has been taken up to allow crystallization or "graining" of the surface layer.

Within certain ranges of relative humidity, candies of any given composition can develop stickiness without ever graining. For example, at

TABLE 14

CHANGES IN HARD CANDIES STORED FOR 30 DAYS AT 70 PER CENT RELATIVE HUMIDITY[1]

Formula, 100 Parts Sucrose With	Average Thickness of Grained Layer Mm.	Average Weight Increase Per cent
90 parts of corn syrup[2]	∼6.5	∼1.35
65 parts of corn syrup	∼5.0	∼1.55
45 parts of corn syrup	∼3.5	∼1.8
15 to 20 parts of invert syrup	∼5.5	>2.6
30 parts of invert syrup	∼7.0	∼1.65

[1] From Heiss (1959).
[2] Corn syrup of 43.9 dextrose equivalent was used in this experiment.

35 per cent R. H. only those candies made with substantial quantities of added invert sugar will grain. At higher relative humidities, the stickiness period is shorter, but the degree of stickiness which is attained is more extreme.

Stickiness generally reaches a maximum twice in the deterioration process—once shortly before the beginning of crystallization and again when the crystal lattice begins to soften. The latter maximum represents a very advanced state of breakdown which is not usually seen under practical conditions of storage. Stickiness tends to decrease with the onset of crystallization.

Stickiness is obviously a texture defect. It gives the candy the appearance and feel of being "pre-licked" in addition to causing the pieces to stick together. Graining or crystal formation dulls and roughens the surface and makes the candy more fragile.

Increases in the content of ash or of caramelized sugars increase the tendency to crystallization, while increasing the amount of corn syrup retards the rate of crystal growth.

Under ordinary conditions of handling, storage, and display, it is essential to pack hard candies in containers having a low moisture-vapor transmission rate. Hermetically sealed tins or glass jars are, of course, ideal.

In these containers, an atmosphere of favorable relative humidity is established by the candy itself and the product will last for a very long time without texture deterioration. Heat-sealed foil-laminates are also very satisfactory. Wrapping of the individual pieces in waxed paper before packaging assists in retarding water vapor absorption.

SUMMARY

Hard candies are the only common products found in the category of glassy-structured foods. They consist of amorphous masses of sugars having very low moisture contents. Consumption is the result of slow solution by saliva and they are usually not meant to be chewed. For these reasons, texture can be considered in this case as a function principally of the surface characteristics, e.g., the roughness or smoothness, and the degree of stickiness. Production methods can have a pronounced effect on texture by causing variations in the number of air bubbles which are present. These bubbles may be perceived as sharp particles.

Hard candies are composed of sucrose, invert sugar, and corn syrup, in addition to minor amounts of other ingredients. The relative proportions of these components determine some of the textural characteristics, such as toughness or brittleness, and influence the rapidity and kind of storage deterioration which will occur under unfavorable conditions.

When hard candies are placed in an atmosphere of higher relative humidity than their equilibrium value (40 per cent R. H. or less), moisture will be absorbed. The initial effect observed is the formation of a sticky surface layer. If the relative humidity is high enough for a sufficiently long period of time, stickiness gives way to graininess as the sucrose begins to crystallize. In extreme cases, the entire piece of candy will degenerate to a fragile agglomeration of crystals. Storage deterioration can be retarded by packaging the candy in containers made of materials having a high resistance to moisture vapor transmission.

BIBLIOGRAPHY

ALIKONIS, J. J. 1955. Carbohydrates in confections. Advances in Chem. Ser. No. 12, 57–63.

BARRET, C. D. 1961. Temperature and relative humidity determine candy shelf life. Food Processing 22, No. 4, 46–47.

BOHN, R. T., and JUNK, W. R. 1960. Sugars. In Bakery Technology and Engineering. Edited by S. A. Matz. Avi Publishing Co., Westport, Conn.

CRAMER, A. B. 1950. Some problems in the manufacture of hard candy. Food Technol. 4, 400–403.

CRAMER, A. B. 1961. Personal communication.

HARVEY, H. G. 1953. The rheology of certain miscellaneous food products. In Foodstuffs: Their Plasticity, Fluidity, and Consistency. Edited by G. W. Scott Blair. Interscience Publishers, New York.

HEISS, R. 1959. Prevention of stickiness and graining in stored hard candies. Food Technol. *13*, 433–440.

KATZ, R., COLLINS, N. D., and CARDWELL, A. B. 1961. Hardness and moisture content of wheat kernels. Cereal Chem. *38*, 364–368.

LIPSCOMB, A. G. Rheology of candy syrup and boiled candies. Fette Seifen Anstrichmittel 58, 875–879.

MAKOWER, B., and DYE, W. B. 1956. Equilibrium moisture content and crystallization of amorphous sucrose and glucose. J. Agr. Food Chem. *4*, 72–77.

MALLOWS, J. H. 1960. Foods of simple structure (sugar syrups, boiled sweets). Soc. Chem. Ind. Monograph 7, 10–13.

MARTIN, L. F. 1955. Sugar in confectionery. Advances in Chem. Ser. No. 12, 64–69.

PALMER, K. J., DYE, W. B., and BLACK, D. 1956. X-ray diffractometer and microscopic investigation of crystallization of amorphous sucrose. J. Agr. Food Chem. *4*, 77–81.

PECKHAM, G. T., JR. 1955. Starch hydrolyzates in the food industry. Advances in Chem. Ser. No. 12, 43–48.

WOODROOF, J. G. 1955. Freezing candies. Georgia Agr. Expt. Stas. Bull. *278*.

Foams and Sponges

INTRODUCTION

The category of foams and sponges is intended to encompass those foods in which small bubbles of gas normally constitute a major portion of the volume. These products consist of a discontinuous phase of gases (usually those of air) and a continuous liquid or solid phase which supports and maintains the structure. Some of the most important types of foods included in the present category are meringues, mousses, souffles, whipped toppings such as whipped cream, chiffon or "fluff" desserts, marshmallow, and leavened bakery products. The last group is probably more important economically than all of the others combined.

Texture in food foams and sponges is, to a considerable extent, a function of the rheological properties of the plastic substances making up the bubble walls. When these properties have a more general importance, they are discussed in one of the preceding chapters. When the system occurs in foods principally as a foam or sponge, as in the case of the starch-and-gluten gel making up the vesicle walls of baked products, it will be discussed in this chapter.

FOAMS AND SPONGES

Perceived Texture

Some of the factors affecting the texture of foams and sponges as perceived in the mouth are the uniformity of the vesicle size, the average size of the vesicles, the thickness of the walls of the vesicles, the shape of the vesicles, the rheological properties of the substance making up the walls, and the size of the ingested piece. In many products, the net effect of most of these factors can be considered negligible, the texture being found to be closely correlated with the physical characteristics of the continuous phase and the size of the vesicles.

Although the gas in the cells is always virtually at atmospheric pressure when the finished product is at rest, the cell walls may restrict egress of the gas sufficiently to cause positive pressure of detectable magnitude during biting and chewing. This occurs, for example, when bread and marshmallow are eaten, but is of no significance in such foods as layer cakes.

The rheological properties of the substance comprising the vesicle walls may be quite different from those of the substance *en masse*. The extension and orderly orientation of microfibrils and large molecules which occur in films generally gives greater elasticity or rigidity to the structure. A foam may be considered to be somewhat analogous to a crystal in that maximum use is made of inter-particle bonding to yield relatively rigid solid structures from weaker agglomerates. Foams are generally firmer when the cell size is uniform, losing strength as the distribution of cell size becomes wider. They also tend to gain strength and stability as the vesicles become smaller, down to the point at which the necessary wall thickness is too small to allow the material to form a coherent film.

Most bubble agglomerates are grossly non-uniform even in the best examples obtainable in practice. The usual trend is to the formation of layers of consolidated and collapsed bubbles around the outside of a piece of the food, so as to yield a crust as in marshmallow and most baked products. The crust is always much different in textural qualities than the rest of the piece and can greatly influence the acceptability of the food. If it is sufficiently non-typical, it can cause rejection of the piece by the majority of consumers.

The phenomena involved in bubble formation, bubble stability, and other properties of foams important in texture determination are part of that division of physics called bubble mechanics.

BUBBLE MECHANICS

Handleman *et al.* (1961) have provided an excellent discussion of bubble mechanics in cake batters while the texts of Bikerman (1953) and Adamson (1960) furnish a more general treatment of this subject. Much of the following analysis is derived from these three sources.

It can be shown that spontaneous initiation of bubbles in homogeneous liquids is very unlikely. One proof of this statement is based on the equation for the pressure within a bubble (in excess of the pressure exerted by the liquid), $P_1 = 2\gamma/R$, where gamma is the surface tension and R is the radius of the bubble. The dissolved gas of the solution, being in equilibrium with gas at a pressure, P_2, would lose gas to a bubble in which the partial pressure of the gas is less than P_2 and would take up gas from bubbles in which the partial pressure is more than P_2. Therefore, it is clear that bubbles can grow only when $2\gamma/R$ is less than P. The surface tension remaining the same, P_2 becomes less likely to be reached as R becomes smaller. Thus, for infinitely small values of R, only infinitely large values of P_2 will permit the formation of bubbles.

It is a matter of common observation that copious bubbles can form in apparently homogeneous liquids. The usual explanation is that these liquids contain microscopic gas bubbles, or dust or other solid particles with entrained gas, which act as nuclei for bubble formation. These dis-

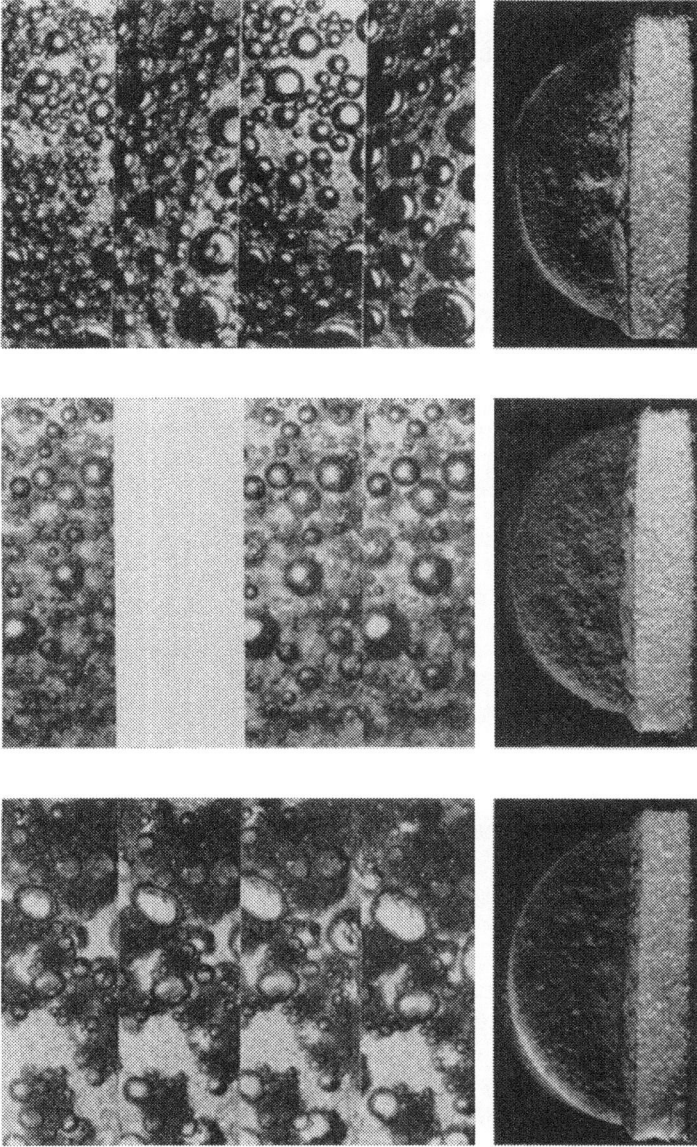

FIG. 25. GROWTH OF BUBBLES IN CAKE BATTERS

Strips show relative bubble size. Standing time increases from top to bottom.

continuities provide foci for the evolution of gas in chemically leavened doughs and batters and in boiling liquids. Foams can also be formed by the injection of gas through small orifices from outside sources or by mechanical agitation. In the latter case, bubbles are drawn into the material by the paddle and are trapped on the surface as the plastic material folds in upon itself to form cavities.

In viscous systems undergoing agitation by any method, the bubbles in the center of the mass tend to become subdivided and to approach a uniform size. In systems at rest, large bubbles tend to increase in size at

TABLE 15

PHYSICAL CHARACTERISTICS OF BATTERS AND FRACTIONS AND EVALUATIONS OF CAKES PREPARED FROM THEM[1]

Variant Formula (Shortening)	Vis-cosity[2]	Yield Value[3]	Surface Tension of Aqueous Fluid (Bubble Wall)	Inter-facial Tension (Fat vs. Water)	Batter Specific Gravity, Gm./Ml.	Cake Specific Volume, Ml./Gm.
Unemulsified						
Whole egg	37	10.5	55	5.6	0.93	2.7
White	72	26	69	12.9	0.93	2.4
White + lecithin	39	2	52	2.3	1.06	2.3
Mono- and diglyceride						
Whole egg	70	39	46	0.7	0.92	3.1
White	83	28	50	0.7	0.94	2.9
White + lecithin	56	10	45	0.6	0.94	2.8
High aerating						
Whole egg	67	32	45	0.5	0.77	3.2
White	160	75	55	0.8	0.75	2.7
White + lecithin	152	85	64	0.6	0.75	3.1
Standard error of measurement	$s \leq 2$	$s \leq 4$	$s \leq 1.5$	$s \leq 10\%$	$s \leq 0.02$	$s \leq 0.03$

[1] Adapted from Handleman et al. (1961).
[2] Centipoises at a low rate of shear of two per second.
[3] In dynes per sq. cm. Yield value calculated from the residual force on a metal plate vertically suspended in a test batter after displacement of the plate and a five-minute relaxation of the batter.

the expense of the smaller bubbles. This conclusion follows from the equation quoted previously, $2\gamma/R = P$, which shows that higher pressures are present in the smaller bubbles. The more uniform the bubbles are, the more stable the foams will be. However, Handleman et al. (1961) point out that apparent exceptions to this rule exist in polydisperse systems containing bubbles completely surrounded by compounds in which the gas that saturates the aqueous phase is not soluble. An example is cake batter made with poorly dispersible shortening. Here the gas bubbles are enveloped in thick layers of fat which inhibit the diffusion of carbon dioxide. Fig. 25 illustrates the growth of bubbles in cake batters containing bubbles surrounded by fat layers of different thicknesses. The cakes which were prepared from these batters demonstrate the effect of bubble size on total retained gas (cake volume).

In systems with uniform bubbles, the system having the smallest bubbles will tend to be more stable. The stability has been empirically estimated to be inversely proportional either to the diameter of the bubbles or to the square of their diameter.

Foams are spontaneously destroyed by collapse and drainage. Collapse results from rupture of the bubbles at the surface or elsewhere, with consolidation of gas spaces or loss of gas to the atmosphere. Collapse generally causes the remaining bubble walls to become thicker. Drainage is the thinning of the bubble walls resulting from the outflow of fluid under the influence of gravity.

Although many phenomena with conflicting effects result from temperature changes in foams, these systems are usually more stable at the lower temperatures.

Substances which affect interfacial tension can exert pronounced effects on foam structure, but the gross results of these changes cannot always be predicted (see Table 15).

Foam Preparation

Preparation methods for food foams and sponges always involve two stages, an aerating step in which a liquid of relatively low viscosity is filled with bubbles by whipping, by injection of gas, or by evolution of a dissolved gas, and a fixing phase which gives the product sufficient rigidity to withstand handling and transportation. The aerating stage can usually be divided into two phenomena, which often overlap or even occur more or less simultaneously. These are the introduction into the fluid of bubbles having widely varying dimensions, and a reduction process in which these bubbles are subdivided until they reach a smaller and more nearly uniform size. In the preparation of bread doughs by the sponge method these two stages are especially distinct. The initial mixing and fermenting yields a plastic mass full of bubbles ranging in size from the very minute to the very large. The second mixing plus the subsequent processing steps such as dividing and rounding partially collapse the large bubbles, subdividing them by pinching the walls together, and cause some of the smaller bubbles to coalesce. In the preparation of sugar-gelatin foams or egg albumen foams, these stages overlap and some new bubbles are introduced even in the final stage of whipping. As the foam becomes stiffer, however, air is not trapped as efficiently as it is when the mass has lower viscosity.

The fixing phase involves a firming or hardening of the continuous phase as a result either of denaturation of the protein component, or of a hardening due to drying or cooling of a plastic mass. Denaturation may occur either as a result of thermal treatment, as in baking, or by surface energy absorption, as in the case of some egg white foams. Fixation by drying and cooling is a common phenomenon in marshmallow manufacture. Rather small losses or translocations of moisture in this product can cause a considerable degree of firming due to the high level of solids originally present.

Surface denaturation is an important texture-affecting process occurring in all foams in which protein helps form the supporting structure. Denaturation of proteins resulting from absorption of surface energy is exactly analogous to heat denaturation. It cannot be reversed, e.g., a collapsed foam of beaten egg white cannot be redissolved by gentle methods. The extent of this denaturation seems to be related to the maximum surface area attained during foaming and to the energy absorbed from the beating instrument. Loss of volume observed in egg white foams when they are whipped excessively is due to a continuing denaturation of the albumen with consequent loss of supporting ability. Uniform thin-but-strong films seem to require an ordered array of like molecules. Denaturation allows the protein molecules to assume a random configuration which is not conducive to film formation.

Angel food cake presents an interesting combination of problems in foam aeration. The foam for this product is prepared by whipping an egg albumen and sugar mixture. The mixture is capable of reaching a high degree of aeration, but it will shrink excessively and exude much liquid when it is baked. Addition of flour to improve the strength of the supporting structure reduces the foaming capacity because of the lipids present in the flour (egg albumen foams are very sensitive to traces of fats). The problem is solved by whipping the albumen until maximum volume is achieved and then folding in the flour. Some loss in volume and rigidity always occurs during the last mixing step and the loss is directly related to the amount of agitation given to the flour-foam system. Reducing the amount of fat present in the flour by extracting it with ether or chloroform produces a material which can be beaten up with the albumen into a foam almost as light and rigid as the albumen foam which does not contain flour. However, upon baking, these foams tend to collapse because the expansion in the oven breaks the more rigid cell walls.

Meringues (egg-sugar foams which do not contain flour) undergo no appreciable expansion in the oven. Form, volume, and most of the desirable texture characteristics are due to the rigid crust formed during baking. The preferred technique is to set this crust by heat before the underlying mass of foam receives enough energy to denature the meringue completely which consequently results in considerable shrinkage and "weeping."

Marshmallows are whipped mixtures of heated corn syrup, sucrose, and water, usually with additives such as invert sugar, glycerin, waxy maize starch, sorbitol, gelatin, albumen, or combinations of these. When the foam has reached its desired density, it is shaped by depositing it into cavities made in cornstarch. Setting-up occurs as a result of cooling, drying, and the forming of a layer of collapsed cells surrounding the outside of the piece. Correct choice of gelatin is important in determining the texture of mashmallow. For making low density (40 oz./gal.) marshmallow to be deposited on cookies, Bronson (1951) recommended gelatin of 175 to 225 bloom strength, while gelatin of about 225 bloom strength is more suitable for ordinary cast white marshmallow where strength is required for handling and packing stability. Light fluffy marshmallow novelties need gelatin of lower bloom strength, e.g., 75 to 125.

On the other hand, Gorfinkle (1961) recommended use of appropriate quantities of the high bloom gelatins selected for texturizing qualities for all of the varieties of marshmallow. He states that high-body gelatins tend to make whiter, more silvery marshmallow. They set faster and have less crust. Gorfinkle also emphasizes that gelatins of the same bloom strength may have different bodying power, some making tougher, and others more tender, marshmallow.

Mallows (1960) lists the most common texture defects of marshmallow as: (1) Graining, most often a result of improper proportions of the different sugars; (2) Toughness, which is affected by the moisture content and the amount and bloom strength of the gelatin; (3) Consolidation, due to high temperature storage; and (4) Crustiness, which is apparently a function of the moisture content of the starch used for molding.

Chiffon- or "fluff"-type desserts are usually prepared by whipping gelatin solutions flavored with sugar and fruit juices. Frequently whipped cream is folded into the mass. Setting-up is accomplished by chilling. Texture is a function of the amount of air whipped into the mass and the amount and quality of the gelatin. Another class of products recently developed is made from fruit purées mixed with small percentages of distilled monoglycerides and beaten. Overruns as high as 1,000 per cent are said to be possible. The mechanism of this process is not clear (Neu and Lee 1961), although the increase in susceptibility to aeration is undoubtedly due to a decrease in the surface tension of the aqueous phase, as has been found by these authors in a study of whipped fruit toppings.

Measurement of Texture in Foams and Sponges

Far more work has been done on texture-determination on leavened bakery products than on any other class of food foams and sponges. The effect on consumer acceptability of relatively small texture changes in bread and cake has long been recognized and the economic significance of these effects has stimulated many investigations. On the other hand, objective texture measurements (other than the texture-related quality of density) on marshmallows and the like have not been described. However, Peter and Bell (1930) devised a method for measuring the stability of whey-protein foams. The foam was prepared under standard conditions. The rate of penetration of a standardized glass tube or penetrometer was taken as an indication of the "stability" of the foam. The tube had an inside diameter of 1.25 cm. and was 42.25 cm. in length. At the upper end, 12 cm. were drawn out to a diameter of 0.4 cm. The weight was 37.5 gm. The foam was transferred to a 250 ml. glass cylinder and shaken down. The penetrometer, which passed through a glass guide to insure a perpendicular fall, was adjusted to touch the foam surface and then released. The initial drop was said to be indicative of the initial strength of the foam, while the fall in cms. at 10- and 20-minute intervals was a measurement of the foam stability. This procedure does not seem to have achieved general acceptance.

Measurements of the texture characteristics of leavened bakery products have usually taken the form of determinations of the resistance to compression of the crumb. The crust is usually not included in the test and resistance to shear or to breaking is infrequently measured although Cornford and Coppock (1950) described a simple instrument for the latter purpose. In this device, a bread sample 1 by 5 cm. in cross section is stretched between two clamps and extended by gradually increasing forces until it ruptures.

Cornford and Coppock also described an instrument for measuring crumb stickiness in the publication quoted above. They first pushed a brass plate down into a slice of baked product and then measured the force required to remove it after a two minute period. They stated that stickiness varies widely but is so sensitive to minor changes in the baking procedure that measurements are of little practical value. Rönnebeck (1957) described a weight adhesion meter, ascribed to Kulman, which he used for rating the stickiness of bread crumb.

The density of baked products is related to the texture, and the measurement of volume is one of the standard quality tests performed in all cereal laboratories. Bread and similar products are measured by rapeseed displacement methods. Occasionally, pearled barley or other seeds are used. Some error is introduced into the measurement by the compression caused by the weight of the seeds lying above the product. In the case of freshly baked products measured in volumeters provided with a long graduated column above a calibrated chamber, the errors may be rather high. Compression errors are reduced when the test chambers are but slightly higher than the loaf, but other more serious errors involving filling and leveling procedure are apt to be introduced in these methods. It is unfortunate that light plastic spheres of uniform diameter are not available for use in volumeters.

The volumes of cakes and biscuits can also be determined by rapeseed displacement, but these products are frequently measured in terms of height at some specific point or points. The approximate volume can then be calculated or read from a table. If the density is to be expressed as specific volume, it is important to use the weight of the baked product in the calculation. Often technicians will use the weight of the batter or dough and the volume of the baked product in the calculation. It should be obvious that such a procedure cannot yield specific volume, density, or any other figure having real dimensions.

Several instruments intended for measuring the compressibility of bread and cake crumb have been described in the literature. These were generally designed to measure either the amount of deformation under a fixed load applied for a given time, or the load required to produce a given deformation in a given time. The elastic recovery has also been measured by some investigators. It is outside the scope of this chapter to review all of these devices, but it may be worthwhile to consider a few of the more typical examples.

Cornford and Coppock (1950) compressed a sample exactly 3.2 cm. in diameter and about one cm. thick between two brass plates by adding weights in 50 gram increments to the top plate. The weight required to compress the sample to half of its original thickness was recorded. This instrument represents the "compressimeter" in its simplest possible form.

Flat plates, either circular or square, are the most common compressing means, but penetrating cones have also been used to measure crumb firmness. Obviously, the two forms of force application do not yield equivalent results. The Precision Penetrometer is one of the pieces of apparatus which has been employed to measure resistance to penetration. It is applicable only to firm crumb, such as that of cake. Evidently, hemispherical plunger tips have not been used in this field.

Bailey (1930) described an apparatus consisting of a simple box and plunger which was lowered on a piece of bread measuring $1^1/_2$ by $1^1/_2$ by 2 inches. The amount of compression was measured after one minute of stress. This instrument represents one of the earlier efforts in the field.

Bradley (1949) used the Morison penetrometer to measure the extent of compression under a fixed load. This device consists of a plunger and a seven pound weight which depresses the plunger.

The device most commonly used in this country to measure the firmness of bread is the Baker Compressimeter. Among the workers who have described this instrument are Platt and Powers (1940). The Baker apparatus utilizes a motor to apply stress at a uniform rate through a plunger to a piece of bread crumb. The strain is indicated on a graduated scale. The size and shape of the bread piece and the point at which the strain is recorded has varied from investigator to investigator. In a recent study, Bechtel (1959) used slices of center crumb measuring 1 by 1 by $^1/_2$ in. and used a compression of 2.5 mm. Crossland and Favor (1950) indicated that it is not desirable to use whole slices as test pieces because of the effect of shear at the perimeter of the plunger. They suggested the use of prisms smaller than the plunger.

In another recent study, Ofelt *et al.* (1958) used a modified Bloom gelometer. Firmness was expressed as the grams of shot required to cause a disc one inch in diameter to depress the bread four mm. Slices were one-half inch in thickness.

There is little doubt that accurate compressimeter measurements are highly correlated with panel assessments of bread firmness.

SUMMARY

Foams include those foods in which small bubbles of gas constitute a major portion of the volume. Sponges are a type of foam distinguished by their greater firmness. Breads, cakes, marshmallow, whipped cream, and fluff-type desserts are representatives of these classes.

Texture in foams and sponges is a function of the size, shape, and size distribution of the bubbles as well as the rheological properties of the film-forming component. The physical characteristics of the continuous phase may differ markedly from those of the same substance in unaerated form.

Preparation methods for food foams can generally be divided into two stages, an aerating stage in which bubbles are brought into suspension and then subdivided, and a fixing step in which the supporting film is made more rigid to give a greater degree of permanence to the foam. Aeration can be accomplished by injecting gas, by evolving gas from a supersaturated solution, or by beating air into a fluid. Fixing is usually the result of heat denaturation, drying, cooling, or surface denaturation. Foams with uniform-sized small bubbles are more stable and often firmer in texture than foams made up of larger bubbles or bubbles falling in a wider range of sizes.

Objective evaluation of the texture of foams and sponges has usually taken the form of measuring the resistance of the material to compression. Use of a fixed deformation is thought to be superior to use of a fixed force. The texture-related characteristic of specific volume is also much used as a quality indication. Resistance to shear, to breaking, or to penetration by sharp plungers are infrequently used as indicators of texture.

BIBLIOGRAPHY

ADAMSON, A. W. 1960. Physical Chemistry of Surfaces. Interscience Publishers, New York.

AMSEL, O. 1942. Foams for fighting mineral oil blazes. Oel u. Kohle 38, 293–304.

BAILEY, L. H. 1930. A simple apparatus for measuring the compressibility of baked products. Cereal Chem. 7, 340–345.

BARRET, C. D. 1961. Combine humectants to extend candy shelf life. Food Processing 22, No. 5, 67–68.

BECHTEL, W. G. 1959. Staling studies of bread made with flour fractions. V. Effect of a heat-stable amylase and a cross-linked starch. Cereal Chem. 36, 368–377.

BECHTEL, W. G., and KULP, K. 1960. Freezing, defrosting, and frozen preservation of cake doughnuts and yeast-raised doughnuts. Food Technol. 14, 391–394.

BECHTEL, W. G., and MEISNER, D. F. 1952. Present status of the theory of bread staling. Bakers Digest 26, No. 1, 5–7.

BIKERMAN, J. J. 1953. Foams: Theory and Industrial Applications. Reinhold Publishing Corp., New York.

BLINC, M. 1956. Some observations concerning the effect of beta-amylase on the staling of bread. Brot u. Gebäck 10, 249–252.

BOURNE, E. J., TIFFIN, A. I., and WEIGEL, H. 1960. Interaction of starch with sucrose stearates and other antistaling agents. J. Sci. Food Agr. 11, 101–109.

BOURREAU, M. 1959. Securing and maintaining softness in sponge cakes by use of sorbitol. Bull. Union des Fabricants de Biscuits 2, 1–8.

BRADLEY, W. B. 1949. Bread softness and bread quality. Bakers Digest 23, No. 1, 5–7.

BRADLEY, W. B., and THOMPSON, J. B. 1950. The effect of crust on changes in crumbliness and compressibility of bread crumb during staling. Cereal Chem. 27, 331–335.

BRONSON, W. 1951. Technology and utilization of gelatin. Food Technol. 5, 51–54.

CATHCART, W. 1940. Review of progress in research on bread staling. Cereal Chem. 17, 100–121.

CLUSKEY, J. E., TAYLOR, N. W., and SENTI, F. R. 1959. Relation of the rigidity of flour, starch, and gluten gels to bread staling. Cereal Chem. 36, 236–246.

COPPOCK, J. B. M., and CORNFORD, S. J. 1960. Texture in bread and flour confectionery. Soc. Chem. Ind. Monograph 7, 64–74.

CORNFORD, S. J., and COPPOCK, J. B. M. 1950. Measuring the physical characteristics of bread. Research 3, 558–562.

CROSSLAND, L. B., and FAVOR, H. H. 1950. A study of the effects of various techniques on the measurement of the firmness of bread by the Baker compressimeter. Cereal Chem. 27, 15–25.

FARRAND, E. A. 1957. Observations on the aerating properties of egg and the mass production of cake. Chem. Ind. 1957, 500–505.

GEDDES, W. F., and BICE, C. W. 1946. The Role of Starch in Bread Staling. Quartermaster Corps Report QMC 17–10.

GORFINKLE, W. I. 1959. Extruded marshmallows. Mfg. Confectioner 39, No. 5, 35–36.

GORFINKLE, W. I. 1961. Private communication.

GREUP, D. H., and HINTZER, H. M. R. 1953. Cereals. *In* Foodstuffs: Their Plasticity, Fluidity, and Consistency. Edited by G. W. Scott Blair. Interscience Publishers, New York.

GUNTHER, K. 1959. Protein in candy. Mfg. Confectioner 39, No. 9, 27, 30–32.

HAMPEL, G. 1957A. Estimation of the crumb characteristics with the aid of the Panimeter. Brot. u. Gebäck 11, 45–49.

HAMPEL, G., 1957B. The development of staling and the degree of swelling of starch. Getreide u. Mehl 7, 17–22.

HANDELMAN, A. R., CONN, J. F., and LYONS, J. W. 1961. Bubble mechanics in thick foams and their effects on cake quality. Cereal Chem. 38, 294–305.

JONGH, G. 1961. The formation of dough and bread structures. I. The ability of starch to form structures, and the improving effect of glyceryl monostearate. Cereal Chem. 38, 140–152.

KEYWORTH, C. M. 1952. Yeast-raised baked goods and flour confectionery. British Pat. 808,069. May 22.

KULP, K., PONTE, J. G., JR., and BECHTEL, W. G. 1959. Some factors that affect the staling of white and yellow layer cakes. Cereal Chem. 36, 228–236.

LANCASTER, E. B., and ANDERSON, R. A. 1959. Consistency measurements on batters, doughs, and pastes. Cereal Chem. 36, 420–430.

LEON, S. I. 1958. The manufacture of marshmallows. Food 27, 257–260.

MALLOWS, J. H. 1960. Foams in confectionery with special reference to marshmallows. Soc. Chem. Ind. Monograph 7, 20–23.

NEU, G. D., and LEE, L. J. 1961. From fruit sauces, purées, juice concentrates—whipped dessert toppings. Food Processing 22, No. 5, 58–61.

NOZNICK, P. P., and GEDDES, W. F. 1943. Application of the Baker compressimeter to cake studies. Cereal Chem. 20, 463–477.

OFELT, C. W., MACMASTERS, M. M., LANCASTER, E. B., and SENTI, F. R. 1958. Effect on crumb firmness. I. Mono- and diglycerides. Cereal Chem. 35, 137–141.

OFELT, C. W., MEHLTRETTER, C. L., MACMASTERS, M. M., OTEY, F. H., and SENTI, F. R. 1958. Effect on crumb firmness. II. Action of additives in relation to their chemical structure. Cereal Chem. 35, 142–145.

PENCE, J. W., and HAMAMOTO, M. 1959. Studies on the freezing and defrosting of cakes. Food Technol. 13, 99–106.

PENCE, J. W., and STANDRIDGE, N. N. 1958. Effects of storage temperature on firming of cake crumb. Cereal Chem. 35, 57–65.

PETER, P. N., and BELL, R. W. 1930. Normal and modified foaming properties of whey-protein and egg-albumen solutions. Ind. Eng. Chem. 22, 1124–1128.

PLATT, W. 1930. Staling of bread. Cereal Chem. 7, 1–34.

PLATT, W., and KRATZ, P. D. 1933. Measuring and recording some characteristics of test sponge cakes. Cereal Chem. 10, 73–90.

PLATT, W., and POWERS, R. 1940. Compressibility of bread crumb. Cereal Chem. 17, 60–69.

RÖNNEBECK, H. 1957. Investigations on the development of objective methods for the evaluation of physical properties of bread-crumb. Ernahungsforschung 2, 105–116.

WINTER, J. D., and TRANTANELLA, S. R. 1958. Effect of packaging on keeping quality of frozen bread and cake. Package Engineering 3, No. 8, 32–36.

ZOBEL, H. F., and SENTI, F. R. 1959. The bread staling problem. X-ray diffraction studies on breads containing cross-linked starch and a heat-stable amylase. Cereal Chem. 36, 441–451.

Effects of Processing Methods on Food Texture

CHAPTER 11

Blanching, Cooking, and Canning

INTRODUCTION

All of the processes discussed here involve subjecting food products to high temperatures without intentionally drying them. Dehydration, which is covered in a subsequent chapter, also requires the application of heat, but the effects of high temperatures other than water vaporization are of secondary importance . Blanching, cooking, and canning are processing techniques in which the heat-driven changes are desired and dehydration is avoided if possible. In some of these processes the heat transfer medium is water or steam while in others air convection or radiant energy is used to heat the product but even then drying is substantially prevented and the product is bathed in its own juices. Dehydration and cooking occur simultaneously in such methods as the baking of biscuits and frying of potato chips, and both kinds of changes are necessary for the development of characteristic flavors and textures in these products.

Blanching, cooking, and canning are rather drastic treatments which often cause extensive changes in the texture of foods. Usually these alterations are desirable although they may be, in some cases, unwanted concomitants of another necessary quality. This is especially true of canning, where the high temperatures and relatively long times required for sterilization frequently yield foods with sub-optimal texture characteristics. Cooking often has as its primary goal the modification of texture, as shown by the use in the kitchen of softness or other textural factors as indicators of the end-point of cooking in vegetables, etc.

The plan used in drawing up this chapter is similar to those used for the other chapters in Section III. First, details of the processing methods will be covered briefly, so that readers not familiar with all of the techniques can become acquainted with the conditions existing in them. Next, a discussion of the reactions that occur in the food as a result of the processing conditions and their effects on the basic textural structures is included. Finally, the net results of these reactions in terms of the perceived texture of the end product will be considered along with an outline of the methods used to follow physical changes during heat treatment.

THE PROCESSES

Blanching

The term "blanching" connotes the brief exposure of a raw food to a high temperature. The most common purpose of blanching is the inactivation of enzymes in fresh fruits and vegetables so that enzymatic reactions leading to changes in color, flavor, or texture will not occur during subsequent holding or processing steps. It is an essential part of the dehydration and freezing techniques used for many food products and is often applied to raw materials for canning when they are to be held for considerable lengths of time at moderate temperatures before retorting. Tressler and Evers (1957) give an extensive discussion of the importance of blanching in the processing of frozen foods. Of the three types of heat processing described in this chapter, blanching is the mildest in terms of the changes caused in the food material.

In practice, blanching can be accomplished by submerging the food pieces in a hot water bath, by drenching them with hot water, by passing saturated steam through a container loosely packed with the pieces, by blowing hot air over them ("hot wind blanching"), by exposing them to infrared radiation, or by using combinations of these methods. Blanching by steam under pressure, or by hot sugar or acid solutions have been suggested but are rarely used. The characteristic texture and flavor changes associated with cooking are avoided as much as possible. Often only the outer layers of the pieces are heated to temperatures high enough to inactivate the catalytic proteins, since oxidative (especially browning) and other changes may occur at a much slower rate in the interior. Rather than depending upon carefully controlled temperatures applied for fairly long periods of time, most blanching procedures rely on a brief exposure to a heat transfer medium at a relatively high temperature. Times of 1 to 5 minutes exposure to temperatures near the boiling point can be considered representative although temperatures of 180°F. and even lower are being used. Obviously, piece size and shape, the amount of agitation or rapidity of circulation of the heat transfer medium, and other factors besides temperature and time of exposure affect the extent to which heat-induced changes occur.

Cooking

Cooking usually causes much more extensive changes in texture than are observed in blanching. Heating may be accomplished by hot air convection (roasting, baking), immersion in hot water (boiling) or hot fat (deep fat frying), conduction (pan frying), infrared radiation (broiling), micro-wave radiation, or other means. Frequently combinations of these techniques are employed. As a rule, temperatures above 212°F. are not generated in the food except in surface layers since the evaporation of water from the fluids tends to keep the internal tissues below this level. Parts of the surfaces may be nearly completely dehydrated during broiling, frying, or other "dry" cooking methods and much higher temperatures can then occur in the dehydrated regions.

In the preparation of roasts and the like, pronounced temperature differentials exist throughout the cooking process with consequent differences in texture dependent upon the distance from the surface. Small pieces respond more uniformly to cooking processes, particularly if they are immersed in water or fat, and the variations in texture with distance from the surface, if present, are more likely to be due to inherent differences in tissue composition.

Cooking methods could be classified for texture purposes as water immersion methods and other types. Characteristic responses are observed dependent upon whether or not water is used as the cooking medium. The composition of the water is important. "Pure" water can coagulate proteins, e.g., in meats, with a resultant shrinkage, even before heat is applied. Salt water tends to extract more of the soluble proteins from the cells and the interstitial spaces and texture is affected thereby.

Canning or Retorting

Heating at high temperatures under pressure is the third type of processing method considered in this chapter. Canned products which are acidic in reaction, such as tomatoes, are often processed under relatively mild conditions. A temperature of 190° to 200°F. throughout all particles of the food is usually considered sufficient to cause the destruction of acid-tolerant organisms. This may sometimes be accomplished by filling the hot (near 212°F.) food into the cans and sealing without additional heat treatment. The texture changes in these foods are not qualitatively different from those observed in ordinary cooking procedures. However, low acid foods such as meats, beans, and potatoes require processing temperatures of 240° to 250°F. in order to assure adequate destruction of the spores of food spoilage organisms. Texture changes are likely to be pronounced when foods are treated by these methods, and they may, in fact, proceed to the point where undesirable characteristics may appear. The food is almost always immersed in liquid water or surrounded by water vapor. The heat treatment affects the product more or less uniformly throughout the mass. The temperature rise is usually rapid and the high temperature phase is of intermediate duration as compared to blanching and cooking. At the termination of the heating phase the food pieces are found to be immersed in solutions of their diluted cell sap and added solutes such as salt and sugar. Agitation may be applied by tumbling, inverting, or other techniques applied to the outer container, but the effects of these treatments on texture are minor.

Destruction of the Semipermeability of Cell Membranes

The most important early happening in the alteration of texture by heat applied to foods containing whole cells is the destruction of the selective permeability of the cell membranes. This irreversible, all-or-none, phenomenon occurs when the cell reaches the vicinity of 150°F. There is no known method for preventing or even retarding the reaction. Of course, semipermeability can be destroyed by agents other than heat, but in the blanching, cooking, and retorting of raw food pieces this is the principal cause. Even in the milder blanching treatments all or nearly all of the cells are brought to the critical temperature range.

As soon as the selective permeability of the cell is lost, the internal pressure is permanently reduced. Some cell distension may still be present because of hydrophilic but non-diffusible substances in the cell and may contribute to perceived texture. However, crispness and similar textural properties will no longer exist.

Another consequence of the loss of selective permeability is the extrusion of cell sap into the intercellular spaces and vascular system. The gases originally present in these spaces are ultimately displaced by the fluids. If the tissue has been immersed in water, the solution is diluted and more solutes are washed out of the cells. If the tissue is dry-cooked, the proportion of extra-cellular to intracellular solutes will not reach such a low level. All of these occurrences have an effect on texture, but the extent, and even the direction, of the changes are difficult to predict.

Tissues immersed in syrup during cooking and retorting not only exhibit an uptake of carbohydrate by diffusion, but appreciable adsorption on the cell walls and other micro-structures takes place. The net effects of these transfers of solutes is an increased firmness of the food. Similar but less important changes occur in products packed in brine.

Intracellular barriers resulting from selective permeability of membranes are also destroyed by heating, although there is no certainty that this destruction occurs at the same temperature at which the external, limiting membranes of the cytoplasm are inactivated. In any case, these internal barriers often serve to separate enzymes from their substrates in the intact cell. Heat can, therefore, make substrates available for reaction. Such a process may be responsible for initiating the essentially proteolytic phenomenon of autolysis.

Effects of Heating on the Activity of Enzymes

Reaction rates of the enzymes acting on texture-affecting structures can be expected to accelerate as the temperature rises. The net effect of this increase is related to the rate at which the temperature is raised. In most of the usual processing techniques, the temperature is increased so rapidly that the time available for the enzyme to act before it is heat denatured is quite short and no demonstrable effect relatable to the increased speed of reaction occurs. Some blanching methods apparently do allow changes to take place. In blanching prior to dehydration, temperatures near 212°F. are applied to leafy vegetables for about 3 to 6 minutes, to peas, beans, and corn for about 10 to 20 minutes, and to carrots, potatoes, and the like for about 5 to 10 minutes. It has been shown that many commercial samples of carrots, sweet potatoes, parsnips, and rutabagas contain hydrolyzed starch. Potatoes generally do not. However, investigators have demonstrated that gelatinization but not hydrolysis occurs if the temperature throughout the product can be made to reach 212°F. within 30 seconds. If the vegetable is held at 122°F. for a few minutes, the starch may be almost completely hydrolyzed. The situation with respect to potato is apparently due to the rapid inactivation of its particular amylase, which is completely destroyed before the starch is gelatinized.

Pectinesterase is another native enzyme which may be denatured by blanching temperatures. However, the degree of esterification of pectin is decreased by heating in slightly acid or neutral solutions in the absence of enzyme activity. Two conflicting trends may be observed during the blanching of vegetables containing pectinesterase. As the rate of temperature rise or the final temperature is increased a reduction in the amount of de-esterification due to enzyme activity will occur, but this may be accompanied by an increase in hydrolysis due to non-enzymatic de-esterification.

Hoogzand and Doesburg (1961) showed that the degree of esterification of pectic substances in cauliflower was decreased by activation of pectinesterase during low-temperature, long-time, blanching (15 min. at 158°F.). This treatment yielded products having greater firmness than did a high-temperature, short-time, blanch (3 min. at 212°F.). Firmness was enhanced by the addition of calcium salts and acid to the cauliflower blanched by the first method. Several studies have confirmed the beneficial effects of low temperature blanching for green beans and snap beans destined for canning (Dietrich et al. 1959; McConnell 1956; Robinson et al. 1949; Sistrunk 1959; and Van Buren et al. 1960). The data in Table 16 are representative of the sensory changes observed. In addition, an increase in the blanch time from 1.5 to 5 minutes has been shown to decrease the sloughing of skins, a fairly common texture defect. It is very likely that these phenomena occur as a result of the same mechanism operating in cauliflower. A representative study on the relationship be-

TABLE 16

EFFECT OF BLANCHING TIMES AND TEMPERATURES ON SLOUGHING, SPLITTING, AND FIRMNESS
OF TENDER GREEN SNAP BEANS[1]

Blanch Temperature °F.	Blanch Time Min.	Sediment Ml.	Splits Per cent	Firmness[2] Lbs.
150	2	2	0	236
160	2	2	0	242
170	1	2	0	243
170	2	1	0	250+
170	3	2	0	250+
170	4	1	0	250+
180	1	1	0	210
180	2	1	5	184
180	3	2	10	190
180	4	4	5	178
190	1	11	5	144
190	2	12	35	108
190	3	14	30	104
190	4	14	20	111
200	2	17	60	97

[1] Van Buren *et al.* 1960.
[2] Firmness determined by a texturemeter which measured the force required to move a 7 cm. diameter piston to a level 3.8 cm. from the bottom of a No. 303 can.

tween blanching temperature and the sloughing of skins (which is also related to bean firmness) is that of Sistrunk and Cain (1960), summarized in Table 17, which showed that there exists a very significant negative correlation of the percentage of sloughed skins with both the percentages of pectinates-pectates and the percentage of alkali-soluble pectins.

Further data on changes in the pectic materials of vegetables during blanching are given in Table 18.

Bode (1961) described a process for tenderizing and sweetening peas through the action of an added heat-stable amylase which works during the initial stages of retorting.

Denaturation of Structural Proteins

In addition to the indirect effects on texture resulting from denaturation of enzymes which attack the various structural components, there is a more direct effect of heat due to the coagulation of structural proteins. General discussions of protein denaturation have been given by Neurath *et al.* (1944) and Anson (1945). The phenomenon is of greatest importance in meats.

Bramblett *et al.* (1959) stated

There is a lack of information as to when coagulation of the muscle proteins occurs, but it is generally assumed that coagulation takes place between 135° and 167°F. There is probably no specific coagulation temperature since muscle fiber consists of several proteins, each with a different coagulation temperature.

TABLE 17

EFFECT OF BLANCHING CONDITIONS ON TEXTURE-RELATED CHARACTERISTICS OF GREEN BEANS[1]

Time of Blanch, Min.	Temperatures of Blanch Water								Time of Blanch, Mean
	130°F.	140°F.	150°F.	160°F.	170°F.	180°F.	190°F.	200°F.	
Resistance to Shear of Canned Beans, Kg.[2]									
1.5	24.4	28.7	32.6	31.0	29.3	24.6	20.5	19.2	26.3
5	24.1	31.0	32.7	33.4	30.9	24.9	20.1	19.1	27.0
10	31.9	38.7	36.4	36.5	35.2	26.3	21.7	19.4	30.7
Mean	26.8	32.8	33.9	33.6	31.8	25.3	20.8	19.2	....
Pectates-Pectinates, Dry Weight Basis[3]									
1.5	3.252	3.162	3.231	3.472	3.512	3.009	2.245	1.751	2.954
5	3.248	3.608	3.731	4.201	3.831	3.156	2.408	1.932	3.264
10	3.482	3.866	4.167	4.287	4.116	3.472	2.601	2.137	3.516
Mean	3.327	3.545	3.710	3.987	3.820	3.212	2.418	1.940	....
Starch Content, Dry Weight Basis[4]									
1.5	18.11	17.67	16.66	19.42	21.56	22.88	21.97	21.00	19.91
5	17.16	18.03	18.70	17.92	22.45	23.60	23.07	23.66	20.57
10	16.59	17.68	19.78	17.11	21.99	23.39	23.57	22.62	20.34
Mean	17.28	17.79	18.38	18.15	22.00	23.29	22.87	22.42	....

[1] Adapted from Sistrunk and Cain (1960).
[2] Statistical Analysis: Effect of temperatures—HSD:LSD (0.1), 4.5. Interaction—HSD. Effect of times—HSD:LSD (0.1) 2.8.
[3] Statistical Analysis: Effect of temperatures—HSD:LSD (0.1) 0.247. Interaction—NSD. Effect of times—HSD:LSD (0.1) 0.151.
[4] Statistical Analysis: Effect of temperatures—HSD:LSD (0.1) 2.32. Interaction—HSD. Effect of times—NSD.
Note: NSD = No significant difference; LSD = Least significant difference; HSD = One per cent level of significance as computed by the analysis of variance. Consult original paper for further details on the statistical treatment.

TABLE 18

INFLUENCE OF VARIOUS BLANCHING CONDITIONS ON PECTIC SUBSTANCES IN CAULIFLOWER[1]

	Experiment A		Experiment B				Experiment C			
	Total Quantity of Pectic Substances		Water-soluble Pectin		Insoluble Pectin		Water-soluble Pectin		Insoluble Pectin	
Time and Temperature of Blanching	Content, Per cent	Degree of Esterification	Content, Per cent	Degree of Esterification	Content, Per cent	Degree of Esterification	Content, Per cent	Degree of Esterification	Content, Per cent	Degree of Esterification
4 min. 212°F.	0.20	70.3	0.08	57.7	0.22	55.8	0.12	63.3	0.23	50.6
4 min. 194°F. plus 4 min. 212°F.	0.23	69.0	....	....	....	....	0.14	64.1	0.20	48.3
4 min. 176°F. plus 4 min. 212°F.	0.25	66.3	0.12	56.3	0.21	54.8	....	....	....	....
15 min.158°F. plus 4 min. 212°F.	0.19	63.2	0.13	52.7	0.22	43.6	0.10	59.5	0.25	39.3
15 min. 158°F., 1 hr. cooling, 4 min. 212°F.	0.21	62.7	....	....	....	....	....	....	....	....
30 min. 140°F. plus 4 min. 212°F.	....	....	0.08	52.8	0.24	48.4	....	....	....	....

[1] From Hoogzand and Doesburg (1961). Degree of esterification expressed as per cent of the theoretical maximum.

The coagulation temperature is made still more uncertain by the protective effect of associated colloids and ions in natural substances, by the solubilizing effect of pH's far removed from the isoelectric point, and by the dependence of the reaction on time as well as on temperature. However, it is doubtful that many protein-containing substances can be heated to 160°F. or over for any appreciable length of time without suffering extensive denaturation.

Further light on this important problem was shed by the report of Deatherage and Hamm (1960). According to these authors, a mild denaturation of muscle proteins occurs between 86° and 104°F. during the cooking of beef. Strong denaturation starts at 104°F. and continues to 122°F. resulting in the formation of new stable cross-linkages which are not split off by the addition of acid or base. Between 122° and 131°F. the denaturation and the formation of new cross-linkages are continued, but the decrease of acidic groups is delayed. Above 131°F. the denaturation continues, causing a tighter network of protein structure. At about 149°F. the denaturation is almost, but not completely finished. Denaturation processes thus may proceed at much lower temperatures than previously recognized.

The most important effects of heat on the texture of meats result from the denaturation of muscle fiber proteins and a hydrolysis of connective

TABLE 19

EFFECT OF INTERNAL DONENESS TEMPERATURE ON THE TENDERNESS OF PORK LOIN ROASTS[1]

Internal Doneness Temperature	Initial Tenderness[2]	Residual Tenderness[2]	Warner-Bratzler Shear Values
185 °F.	5.6 ± 0.32	5.3 ± 0.34	7.6 ± 0.69
165 °F.	5.8 ± 0.31	5.6 ± 0.30	7.7 ± 0.56
150 °F.	6.3 ± 0.38	6.0 ± 0.42	7.0 ± 0.50
150 °F.[3]	6.8 ± 0.36	6.4 ± 0.28	6.9 ± 0.50

[1] Webb et al. (1961).
[2] Tenderness was rated on a nine-point hedonic scale by six trained judges. Tender extreme was 9, tough extreme was 1.
[3] In this case the internal temperature was maintained at 150°F. for one hour.

tissue. Apparently there is an optimum temperature at an intermediate level which causes muscle fibers to become more tender than do higher or lower temperatures. Bramblett et al. (1959) showed that the tensile strength of muscle fibers heated at 145°F. was less than the tensile strength of raw muscle fibers or of fibers heated at 154°F. Meat cooked at 145°F. was more tender than meat cooked at 154°F. Webb et al. (1961) also demonstrated this effect in pork cooking as shown in Table 19. Sweetman and Mackellar (1954) also emphasized that heat hardens muscle fibers, but less tender cuts of meat require heating at temperatures above 149°F. Extended heating at a well-done temperature of 176°F. is required to soften the connective tissues.

It has been shown by several investigators that protein denaturation of certain enzymes apparently can be reversed. The role of reversal in structural protein denaturation has not been elucidated but it probably is not an important factor in texture determination.

Gelatinization of Starch

The amount of starch contained in fruits and vegetables varies widely according to type and stage of maturity but it is often present in quantities sufficient to affect texture. In potatoes, for example, the distribution and physical status of the starch is the predominant texture-influencing factor. The small amounts of glycogen in muscle tissues probably have no detectable effect on texture.

The different starches gelatinize in characteristic temperature ranges which may be altered somewhat by the pH of the medium, presence of certain ionic species, and the rate of heating. Crafts (1944) observed starch gelatinization microscopically in potato cells and reported that the starch grains expanded rapidly with a bursting action as though they had been retained within membranes that had suddenly broken. The expanding granules soon filled the cells completely with gray, reticulate, translucent masses that almost obscured the cell outlines.

Most blanching procedures probably do not gelatinize all of the starch granules of the cells. Peas, beans, potatoes, and other foodstuffs which contain large amounts of starch often retain their tough and hard texture throughout the blanching process. On the other hand, all ordinary cooking procedures can be expected to completely gelatinize the starch with development of a tenderer, softer texture. Retorting carries the reaction still further with partial breakdown of the microstructures of the starch masses and even some hydrolysis of the molecules. As a result, the suspending liquid may take on some gel-like qualities while the food loses part of its structural integrity and becomes undesirably soft and mushy.

The quantity of water available to the granule may have an effect on the progress of gelatinization. When the food is boiled or cooked by other methods permitting access to a relatively large amount of water, gelatinization and breakdown is limited only by the intensity of the heating step, and, to some extent, the violence of the agitation. When the food is cooked in its own juices, or by some technique involving partial dehydration, complete gelatinization may be prevented. In such cases, the amount of water which has been added can be the limiting factor in texture determination.

Non-enzymatic Hydrolysis of Polymers

Proteins, pectins, starch, and other polymeric molecules are susceptible to hydrolytic scission at high temperatures. The hydrogen ion concentration is a most important factor controlling the rate of such reactions at a given temperature. The formation of gelatin from collagen is an extreme example of what can be accomplished in this regard. The soluble (dispersible) colloid gelatin is manufactured from the completely insoluble collagen of bones, hides, etc. by acid (and sometimes alkaline) extraction and cooking. Doubtless some hydrolysis of the connective tissues of meats occurs during retorting and probably during ordinary roasting and boiling as well. The "mushy" condition of canned meats is a texture fault which is partially due to this effect. In practice, raw meat containing relatively large proportions of connective tissue is selected for canning to offset the softening resulting from the high temperatures necessary for sterilization. It is fortunate that the pH of most canned meat products does not deviate much from neutrality, the level at which hydrolysis may be expected to be at a minimum.

Visser *et al.* (1960) reported that degree of doneness of beef was not appreciably important to the shear value for muscle. Stadelman and Wise (1961) found similar effects in chicken muscle, at least from slightly cooked to done—the results of overcooking not being examined. The latter investigators published data on the effect of different cooking methods on tenderness of chicken meat. Some of these data, chiefly important because of the inclusion of microwave heating in the methods studied, are reproduced in Table 20.

TABLE 20

EFFECT OF HEATING ON SHEAR VALUE OF CHICKEN MEAT[1]

Heat Treatment	Shear Values[2]
1$\frac{1}{2}$ min. microwave	7.6
6 min. steam	11.7
12 min. steam	10.3
7 min. deep-fat fry	12.5
2 min. fat; 5 min. microwave	14.5
5 min. microwave	12.4
None	7.7

[1] Adapted from Stadelman and Wise (1961). Each value is the average of 5 replications.
[2] Kramer shear press values in lbs. per gm. of meat.

As indicated previously, pectins undergo some de-esterification upon heating in neutral or slightly acidic media (also, see Table 21). This change is considerably accelerated in alkaline media, but such solutions are rarely encountered in fruit and vegetable cookery. The basic chain structure of pectic materials seems to be resistant to hydrolytic scission under the conditions normally existing in cookery and canning. Starches can be expected to undergo some breakdown when subjected to rigorous heat treatments. Cellulose may undergo a change which is either a disruption of multi-molecular aggregates or a partial hydrolysis (Simpson

TABLE 21

EFFECT OF STEAM-COOKING ON PECTIC SUBSTANCES OF CARROTS AND PARSNIPS[1]

Fraction of Pectic Substances	Raw Vegetable Per cent		Steamed 20 Min. Per cent		Steamed 45 Min. Per cent[2]	
	Carrots	Parsnips	Carrots	Parsnips	Carrots	Parsnips
Pectin	3.7	4.7	6.0	6.1	8.8	7.9
Protopectin	14.1	10.2	9.0	7.7	3.6	5.7
Pectic acid or pectates	0.8	1.6	1.0	2.0	1.3	2.1
Total pectic substances	18.6	16.4	16.1	15.8	13.7	15.7

[1] From Crafts (1944).
[2] Data expressed as per cent of the dry substance.

and Halliday 1941). Although detailed studies are lacking, lignin is probably resistant to the effects of heat treatments within the normal ranges.

Melting of Lipids.—Most fruits and vegetables contain such small amounts of fat, and it is so tightly bound, that specific textural consequences of lipid melting are hard to distinguish. The melting of lipids and their removal from the food by dripping or flotation has a marked influence on the texture of meats. This change is manifested not only in the tenderness of the meat but in other textural components as well. The overall effects most commonly observed are shrinking, toughening, and, in some cases, crisping. The extreme example is fried bacon, in which the predominant texture change is a result of the removal of fat from the adipose tissue. In lean meats, the melted lipids soften the cooked food by lubricating the fibrils.

Heating of emulsions to high temperatures causes a separation of the phases. Emulsion separation is not of much importance in the texture of foods prepared by the methods being discussed in this chaper, except that attempts to can emulsion-based foods usually meet with failure when a retorting step is included in the process.

In deep fat frying, the hot fat may replace some of the water in cell voids and intercellular spaces, especially in the surface layers of the food. The redistribution is of particular importance in potato chip preparation where large proportions of absorbed fat affects the texture even though the main influence on perceived texture is the crisp network of dehydrated starch and protein.

Evolution of Gases.—The decreased solubility of dissolved gases (especially carbon dioxide) and the increased vapor pressure of water at the temperatures encountered during heat processing can result in the filling of intra- and inter-cellular spaces with gases and the consequent expulsion of fluids from the vascular networks, rupture of cell walls and larger limiting structures, and, finally, consolidation and shrinkage of tissues as the gases decrease in volume during cooling. Gas evolution is the foundation of foam formation, and texture alteration, in the very important classification of bakery foods which are treated in detail in the chapter on foams and sponges.

Obviously, evolution of water vapor is of small importance to texture in blanching. Cooking by "wet" methods likewise limits the contribution of steam to the breakdown of tissues. However, roasting, frying (pan or deep fat) and similar "dry" cooking techniques can be expected to generate sufficient steam to affect the texture.

Measuring Texture Changes During Cooking

Most of the instruments and methods designed to indicate texture quality can be applied to foods at the various stages of heat processing with the same degree of success as accompanies their use on raw products. The greatest number of published reports in this field have dealt with the changes in meat texture.

Bard and Tischer (1951) followed the rate of increase in tenderness of canned beef during heat processing by measuring resistance to shear, drained juice weight, and the moisture content of the meat. They concluded that the moisture content of meat after draining off the juice, and the weight of the drained juice, are comparable criteria for describing the changes in juiciness as affected by thermal processing. Cole *et al.* (1960) found that drip loss was correlated with development of tenderness during roasting of beef.

Asselbergs and Whitaker (1961) developed a simple hydraulic pressure method for determining the free moisture content of cooked meat. In essence this method involved the application of a given pressure to a small sample of ground meat contained in a special test cell. After a set time, the pressure was released and the meat sample re-weighed to determine the amount of fluid which had been expressd from it. A technique for determining drip loss during cooking was developed from this method.

Bramblett *et al.* (1959) rated the increase in tenderness during cooking by measuring the tensile strength of individual muscle fibers with the device of Smith (1957), and by using the Warner-Bratzler shear technique. Significant differences in tenderness of meats cooked at different temperatures were indicated by these methods.

Visser *et. al.* (1960) were interested in determining the effect of degree of doneness on the tenderness and juiciness of beef cooked by two different methods—oven roasting and deep fat cooking. They found that, in general, correlation coefficients for Warner-Bratzler shear values vs. tenderness (panel) scores, and juiciness scores vs. press fluid yields were not significant. Press fluid was determined by pressing ground samples at 4,000 lbs. per sq. in. and measuring the fluid expressed. Stadelman and Wise (1961) used Kramer shear press values as an indication of the effect of cooking conditions on the tenderness of poultry.

BIBLIOGRAPHY

ANSON, M. L. 1945. Protein denaturation and the properties of protein groups. Advances in Protein Chem. 2, 361–386.

ASSELBERGS, E. A., MOHR, W. P., and KEMP, J. G. 1960. Studies on the application of infrared in food processing. Food Technol. 14, 449–453.

ASSELBERGS, E. A., and WHITAKER, J. R. 1961. Determination of water-holding capacity of ground cooked lean meat. Food Technol. 15, 392–394.

BARD, J. C., and TISCHER, R. G. 1951. Objective measurement of changes in beef during heat processing. Food Technol. 5, 296–300.

BODE, H. E. 1961. Enzyme acts as tenderizer. In Practical New Canning and Freezing Methods. Chilton Publications, Philadelphia.

BRAMBLETT, V. D., HOSTETLER, R. L., VAIL, G. E., and DRAUDT, H. N. 1959. Qualities of beef as affected by cooking at very low temperatures for long periods of time. Food Technol. 13, 707–711.

CALDWELL, J. S., and CULPEPPER, C. W. 1943. Snap bean varieties suitable for dehydration. Part 2. Food Packer 24, 363–368.

COLE, J. W., BACKUS, W. R., and ORME, L. E. 1960. Specific gravity as an objective measure of beef eating quality. J. Animal Sci. 19, 167–173.

COVER, S., KING, G. T., and BUTLER, O. D. 1958. Effect of carcass grades and fatness on tenderness of meat from steers of known history. Texas Agr. Expt. Sta. Bull. 889.

CRAFTS, A. S. 1944. Cellular changes in certain fruits and vegetables during blanching and dehydration. Food Research 9, 442–452.

DAVIS, R. B., DE WEESE, D., and GOULD, W. A. 1954. Consistency measurements in tomato purée. Food Technol. 8, 330–334.

DEATHERAGE, F. E., and GARNATZ, G. 1952. A comparative study of tenderness determination by sensory panel and by shear strength measurements. Food Technol. 6, 260–262.

DEATHERAGE, F. E., and HAMM, R. 1960. Influence of freezing and thawing on hydration and charges of the muscle proteins. Food Research 25, 623–629.

DIETRICH, W. C., OLSON, R. L., NUTTING, M., NEUMANN, H. J., and BOGGS, M. M. 1959. Time-temperature tolerance of frozen foods. XVIII. Effect of blanching conditions on color stability of frozen beans. Food Technol. 13, 258–261.

FRANK, S. S., and CIRCLE, S. J. 1959. The use of an isolated soybean protein for non-meat, simulated soybean sausage products, frankfurter and bologna types. Food Technol. 13, 303–313.

GOULD, W. A. 1951. Quality evaluation of fresh frozen, and canned snap beans. Ohio Agr. Expt. Sta. Bull. 701.

HEINZE, P. H., KIRKPATRICK, M. E., and DOCHTERMAN, E. F. 1955. Cooking quality and compositional factors of potatoes of different varieties from several commercial locations. U. S. Dept. Agr. Tech. Bull. 1106.

HOOGZAND, C., and DOESBURG, J. J. 1961. Effect of blanching on texture and pectin of canned cauliflower. Food Technol. 15, 160–163.

HUGHES, R. E., JR., CHICHESTER, C. O., and STERLING, C. 1958. Penetration of maltosaccharides into processed clingstone peaches. Food Technol. 12, 111–115.

LAZAR, M. E., CHAPIN, E. O., and SMITH, G. S. 1961. Dehydro-frozen apples. Recent developments in processing methods. Food Technol. 15, 32–36.

LEONARD, S., LUH, B. S., and MRAK, E. M. 1954. Factors influencing drained weight of canned clingstone peaches. Food Technol. 12, 80–85.

Levine, A. S., Fellers, C. R., and Barton, R. R. 1950. Preservation of Russian caviar by canning. Food Technol. *4*, 15–16.

McColloch, R. J., Nielsen, B. W., and Beavens, E. A. 1950. Factors influencing the quality of tomato paste. II. Pectic changes during processing. Food Technol. *4*, 339–343.

McConnell, J. E. W. 1956. Factors affecting the skin of processed green beans. Confidential Research Report 205-56. National Canners Assoc., San Francisco.

Moyer, J. C., Robinson, W. B., Ransford, J. R., LaBelle, R. L., and Hand, D. B. 1959. Processing conditions affecting the yield of tomato juice. Food Technol. *13*, 270–275.

Neurath, H., Greenstein, J. P., Putnam, F. W., and Erickson, J. O. 1944. The chemistry of protein denaturation. Chem. Revs. *34*, 157–265.

Olson, N. F. 1959. Study of the control of the physical structure of pasteurized process cheese spreads. Dissertation Abstrs. *19*, 2701–2702.

Reeve, R. M., and Neufeld, C. H. H. 1959. Observations on the histology and texture of Elberta peaches from trees of high and low levels of nitrogen nutrition. Food Research *24*, 552–563.

Robinson, W. B., Moyer, J. C., and Kertesz, Z. I. 1949. "Thermal maceration" of plant tissue. Plant Physiol. *24*, 317–319.

Schoman, C. M., Jr., and Ball, C. O. 1961. The effect of oven air temperature, circulation, and pressure on the roasting of top rounds of beef (yield and roasting time). Food Technol. *15*, 133–136.

Sherman, P. 1961A. The water binding capacity of fresh pork. I. The influence of sodium chloride, pyrophosphate, and polyphosphate on water absorption. Food Technol. *15*, 79–87.

Sherman, P. 1961B. The water binding capacity of fresh pork. III. The influence of cooking temperature on the water binding capacity of lean pork. Food Technol. *15*, 90–94.

Simpson, J. I., and Halliday, E. G. 1941. Chemical and histological studies of the disintegration of cell membrane materials in vegetables during cooking. Food Research *6*, 189–206.

Sistrunk, W. A. 1959. Effect of certain field and processing factors on the texture of Blue Lake green beans. Dissertation Abstr. *20*, 983–984.

Sistrunk, W. A., and Cain, R. F. 1960. Chemical and physical changes in green beans during preparation and canning. Food Technol. *14*, 357–362.

Smith, E. E. 1957. Tenderness and chemical changes in muscle fibers during cooking. Master's Thesis, Purdue University.

Stadelman, W. J., and Wise, R. G. 1961. Tenderness of poultry meat. I. Effect of anesthesia, cooking, and irradiation. Food Technol. *15*, 292–294.

Sterling, C. 1959. Drained weight behavior in canned fruit: An interpretation of the role of the cell wall. Food Technol. *13*, 629–634.

Sterling, C., and Chichester, C. O. 1960. Sugar distribution in plant tissues cooked in syrup. Food Research *25*, 157–160.

Strohmaier, L. H. 1953. Studies on the toughness and histology of frozen apricot skins. Food Technol. *7*, 469–473.

Sweetman, M. D., and Mackellar, I. 1954. Food Selection and Preparation. Fourth Edition. John Wiley and Sons, New York.

Tressler, D. K., and Evers, C. F. 1957. The Freezing Preservation of Foods. Vol. I. Avi Publishing Co., Westport, Conn.

Truscott, J. H. L., and Wickson, M. 1956. Fruit maturity as indicated by juice viscosity. Rept. Hort. Prod. Lab., Vineland, Ontario, *1953–1954*, 105–107.

Van Buren, J. P., Moyer, J. C., Wilson, D. E., Robinson, W. B., and Hand, D. B. 1960. Influence of blanching conditions on sloughing, splitting, and firmness of canned snap beans. Food Technol. *14*, 233–236.

Veis, A., and Cohen, J. 1960. Reversible transformation of gelatin to the collagen structure. Nature *186*, 720–721.

VISSER, R. Y., HARRISON, D. L., GOERTZ, G. E., BUNYAN, M., SKELTON, M. M., and MACKINTOSH, D. L. 1960. The effect of degree of doneness on the tenderness and juiciness of beef cooked in the oven and in deep fat. Food Technol. *14*, 193–198.

WANG, H., ANDREWS, F., RASCH, E., DOTY, D. M., and KRAYBILL, H. R. 1953. A histological and histochemical study of beef dehydration. I. Rate of dehydration and structural changes in raw and cooked meat. Food Research *18*, 351–359.

WEBB, N. L., RUTHERFORD, B. E., and WIANT, D. E. 1961. Effect of ionizing radiation on cakes and biscuits made with milled irradiated wheat. Food Technol. *15*, 386–388.

WHITTENBERGER, R. T. 1951. Measuring the firmness of cooked apple tissues. Food Technol. *5*, 17–20.

CHAPTER 12

Freezing

INTRODUCTION

Bringing foods to a temperature which renders solid all of the liquids therein not only has an immediate and obvious effect on the texture of the product but also affects the quality of the food after thawing to an extent dependent upon the freezing rate, the time, temperature, and relative humidity of storage, the conditions of thawing, and the composition of the food.

In general, the faster the rate of freezing and the lower the storage temperature, the better the original texture of the food will be preserved. Rapid thawing at moderate temperatures is also conducive to quality retention. Some foods are relatively insensitive to changes in these conditions while others have extremely critical requirements for proper freezing and storage. One extreme form of behavior is exemplified by fresh tomatoes and lettuce which have their characteristic texture completely destroyed when they are subjected to any known combination of freezing conditions, while the other extreme may be considered to be represented by some of the dehydrated foods such as potato flakes or by fruit powders for infant feeding which have a recommended storage temperature of about 0°F. The latter products contain water in an amount insufficient to form crystals at any temperature however low, i.e., they contain no "free" water.

CHANGES OCCURRING DURING FREEZING

General

The initially observable effects of freezing on texture are due principally to the formation of ice crystals with resultant physical distortion of the cell and its contents, dehydration of hydrophilic colloids, and concentration of soluble solids. In crisp vegetables, for example, these changes are quite destructive to the characteristic texture of the food. In some other products, e.g., muscle tissues, there is usually very little initial change in the sensory qualities, and, in fact, the cooked meats prepared from frozen and refrigerated raw materials may be indistinguishable to the average consumer.

The rate of freezing has a pronounced effect on the size of the crystals. The more rapidly the product is frozen, the smaller are the crystals. The kind and amount of solubles also influence the size of the crystals in addition to lowering the freezing point and making the freezing rate effectively lower under any given set of conditions. The change of water to ice is accompanied by an increase in volume. Pure water at 32°F. increases in volume about 8.6 per cent as it passes from the liquid into the solid state. Consequently, the cells are subjected to distorting and puncturing forces as the ice crystals separate out from the concentrated solutions of cell sap.

Muscle Tissues

Many studies with meat, poultry, and fish have shown that slow freezing does indeed cause the development of a relatively small number of large ice crystals while fast freezing results in the formation of a large number of very small crystals. When rapidly frozen meat is sectioned either longitudinally or across the fibers, little difference from unfrozen muscle can be discerned even under rather high magnification. Definite changes in tissue structure can be seen in meats which have been slowly frozen and these changes appear to be related to the formation of ice crystals both within the cells and in the intercellular spaces.

Tarr (1947) found that most of the texture deterioration which occurs in fish upon freezing, especially if the fish are slowly frozen, is due to a denaturation of myosin, the predominant protein in the muscle. The myosin is rendered insoluble by some chemical or physical change, perhaps dehydration, occurring during freezing. This denaturation occurs fastest between 23° and 30°F. although it also can occur at much lower temperatures. Retardation of this undesirable phenomenon is best accomplished by rapid freezing and storage at a constant low temperature, preferably —4°F. or lower. Earlier workers generally attributed the bad effects on texture of slow freezing to the large ice crystals causing rupture of muscle cells, a spongy appearance, and excessive drip during thawing. Tarr does not rule out the possibility that minor contributions to texture deterioration in frozen fish might be made by these actions of large ice crystals.

Tressler *et al.* (1932) indicate that freezing effects a definite tendering of meat. They employed two different texture measuring devices in the reported study. One was an ordinary tire pressure gage having a blunt probe inserted in it. The puncturing part of the instrument was 2.5 inches long and $5/16$ in. in diameter. The rounded point had a radius of about 0.08 in. The point was pushed eight times through a slab of meat held securely in a press. The second device was a penetrometer of a type commonly used for testing bituminous materials. The plunger was replaced by a needle $1^3/8$ in. long, 0.15 in. in diameter, and about 0.07 in. radius. A 255 gram weight was used to force the instrument into the steak.

Using the two instruments described above, Tressler *et al.* found that the tendering action of quick freezing varied from about 20 to about 50 per cent. This tendering effect is apparently less important in aged beef.

Contrary to the results of some other studies, Deatherage and Hamm (1960) found that quick freezing and thawing of beef resulted in no appreciable denaturation of muscle protein but do cause a very small but significant increase of the water-holding capacity, probably due to the formation of tiny ice crystals inside the cells. These investigators stated that slow freezing causes a significant small decrease of the water-holding capacity of meat, perhaps due to some destruction of the protein structure by the formation of large ice crystals between the cells.

Fruits and Vegetables

Lee *et al.* (1946) studied the development of crystals in the tissues of green beans and peas and found that, as expected, larger crystals developed in the vegetables which were frozen more slowly. However, after cooking stored products, they could not detect any difference in texture between foods given the two treatments. Destruction of tissues in some other fresh foods is more extreme. MacArthur (1945) found that tissues of asparagus, strawberries, and corn were badly torn by internal ice crystals when they were slowly frozen, while the more rapid methods caused less damage. Woolrich and Bartlett (1942) had previously reported that the smallest ice crystals are much larger than the individual cells even when freezing is exceedingly fast. Many cells were found to be contained in one [mass of] ice crystal[s], instead of the reverse as might be expected. These authors indicated that the crystal lattice both inside the cell and in the intercellular spaces was continuous, but that neither tearing nor shearing of the cell walls occurred regardless of the method of freezing employed. They acknowledged that such damage might occur due to crystal formation during slow freezing of some classes of foods.

One of the most extensive studies of ice crystal formation in frozen fruits was that of Woodroof (1938). He gave three reasons for the development of flabbiness or weakness of structure in thawed mature fruits. These were: (1) puncturing of cell walls by ice crystals; (2) withdrawal of more water from the cells into the intercellular spaces than is reabsorbed on thawing; and (3) destruction of the colloidal complex of the cells with reduced turgidity.

According to Woodroof's investigations, cell rupture is very pronounced in mature fruits such as strawberries, raspberries, peaches, cherries, etc. which consist of very thin parenchyma cells with large intercellular spaces. In these fruits, the water constitutes 84 to 89 per cent and the sugars 5 to 10 per cent of the total weight, thus facilitating the development of large ice crystals.

Tressler and Evers (1957) likened the action of freezing on vegetable texture to that of the first stages of cooking. The over-all effect is a wilting which destroys the desired crisp character of raw vegetables—the products become limp and flaccid. If the products are to be cooked after frozen storage, they ordinarily require only about one-half the usual cooking time as a result of the texture changes induced by freezing (and blanching).

The study of Woodroof previously discussed goes into considerable detail regarding the microscopic changes observed in frozen vegetables. In addition to the crushing and rupturing effect of ice crystals which may be a predominant cause of texture change in mature fruits, he describes two other effects operating to produce flaccidity in crisp vegetables: (1) flabbiness results from withdrawal of more water from the cells (into external ice crystals) than can be reabsorbed upon thawing; and (2) turgidity is reduced due to destruction of the colloidal complex of the cells. The first phenomenon is apparently related in part to the degree of cell breakdown, since whole cells in the vegetables could absorb practically all of the original water, although they held some of it rather loosely. The reduction in turgidity is evidently dependent upon the content of starch, sugars, and salts, which exert a protective effect on the colloidal complex.

Eggs

Eggs respond well to freezing, and very large quantities of the frozen variety are sold to food processors, especially to bakers, either as whole, separated, or modified products. The viscosity of yolks and whole eggs are changed slightly by freezing (Pearce and Lavers 1949). These changes can be partially counteracted by adding a large enough concentration of salt, sugars, or glycerol to the eggs before they are processed. At any rate, the changes apparently do not affect the performance of the eggs in such bakery products as pound cakes, especially if the eggs are allowed to age for a sufficient period (about three months) after freezing. Little change occurs in raw egg white upon freezing. It performs about as well as fresh white in such critical products as angel food cakes and meringues. However, cooked egg white becomes rubbery and unpalatable. This alteration is probably due to dehydration and partial denaturation of the proteins responsible for gel formation. Davis et al. (1952) concluded instead that the damage is a result of the mechanical effects of the ice crystals formed. Their explanation is as follows:

During freezing, the water in the elastic gel of the cooked egg white (denatured protein) migrates to increase the size of crystals wherever nuclei are present. As the crystals grow, they penetrate the gel and separate the structure, thus releasing a part of the elastic tension. The migration of the water from within the gel structure, plus the force exerted by the growth of ice crystals and the release of elastic tension by mechanical cleavage, cause the gel structure to contract. That this contraction is largely irreversible is demonstrated by the liquid-filled spaces remaining after thawing. The structure remaining is naturally tougher, since it contains a considerably higher proportion of protein than the original gel. Thus if a gel before freezing contains twelve per cent solids and if a 55 per cent liquid separation occurs during freezing, as occurred in many experiments, then the remaining gel would contain upward to 27 per cent solids. The actual remaining solids will be somewhat lower than 27 per cent because the liquid will contain part of the soluble solids.

Cereal Products

Unbaked doughs and batters are not appreciably affected by freezing. Although the critical component in doughs is a hydrated protein (gluten), the moisture translocation which so adversely affects some other hydrophilic compounds during freezing is apparently not a major factor. Perhaps the forces holding the water molecules to gluten are stronger than those binding water to, for example, the protein molecules in egg albumen. A very important factor governing texture of cakes baked from frozen batters is the stability of the emulsion of shortening in the aqueous phase. Properly formulated and emulsified batters are little affected by freezing.

Some initial loss of carbon dioxide is to be expected in leavened doughs and batters. This compound, like the other soluble materials, becomes increasingly concentrated in the fluid phase with a consequently increasing tendency to be released as gas into the surrounding atmosphere. Yeast-leavened doughs are also damaged initially to a fairly small extent by freezing. In this case, loss of fermentation capacity of the yeast cells is the cause of increased density and impaired texture in the baked product. Although a considerable portion of the yeast cells will usually survive almost any freezing procedure, some of them are always killed and an even larger number have their fermentation capacity reduced by varying amounts. Rapid freezing is desirable because it minimizes osmotic damage resulting from the concentration of solutes to high levels in the fluid surrounding the cells. Damage to yeast cells is greatly reduced if they are not permitted to bud before freezing commences. Doughs frozen before proofing are superior to doughs frozen after proofing.

Cookie doughs contain little or no leavening and have a very high concentration of soluble materials and fats. They respond very well to freezing. Indeed, most investigators have been able to detect no difference between cookies made from fresh dough and those made from the same dough after freezing. Pie doughs are generally not harmed by freezing, so far as texture is concerned. They contain relatively small amounts of water and relatively high amounts of fat and are not dependent upon the existence of an emulsion for proper performance. Accretion of ice crystals in the dough mass by mechanisms previously described has little effect on the baked quality of the dough.

Low moisture baked products such as pie crusts, cookies, crackers, etc. exhibit no detectable textural changes as a result of freezing. Bread, rolls, and doughnuts may show some hardening or toughening on freezing as a result of increased rates of starch retrogradation near the freezing point. If the freezing is performed very rapidly, the total effect may be quite insignificant. Generally, some volume loss (contributing to texture changes) is observed in high moisture baked products and this is probably also related to the starch retrogradation or the staling phenomenon and to moisture loss. Cakes are not affected by slow freezing. Other factors than starch retrogradation appear to dominate in the hardening of cakes.

Confections

Most types of candy can be frozen and thawed without any harm befalling their texture, and some may actually be improved by the process. Woodroof (1955) has published a series of reports on this subject. He found that the principal problem arising out of the freezing process itself is the cracking which occurs in spun candy chips, some chocolate coated nuts, brittles, toffees, and a few types of creams. The defects are principally in the visual area except in the case of spun candy chips where actual shattering may occur with a resultant deterioration of texture. Doubtless this cracking is the result of the reinforcement of previously existing internal tensions by thermal contraction.

Other Products

Oil and water emulsions such as salad dressings, mayonnaise, and the like may be broken down by freezing. The processing forces the water to crystallize and the oil to solidify, disturbing the emulsion and permitting the coalescence of the discontinuous phase when thawing occurs. The damage seems to be more extreme if the freezing and thawing are performed slowly. Tressler and Evers (1957) state that homogenization of these emulsions helps to stabilize them. Emulsifiers such as glycerol monolaurate, monopalmitate, and monostearate are also effective, especially if added prior to a homogenization step.

Most of the pioneer packers of frozen meals encountered the problem of separation or curdling of cream sauces and gravies. Although the drop in acceptability due to curdling may be mostly a result of the change in appearance, a texture defect is also introduced by the change from a smooth and homogenous gel to a watery suspension of rubbery particles.

This phenomenon is the result of retrogradation of the starch (which permanently reduces its solubility), the coalescence of water into relatively large ice crystals from the colloidal size droplets initially present, and the denaturation of egg proteins, if any are present. Changes in the proportions of fat and water, the type of fat, and the processing conditions apparently help very little in overcoming this problem. Stability of

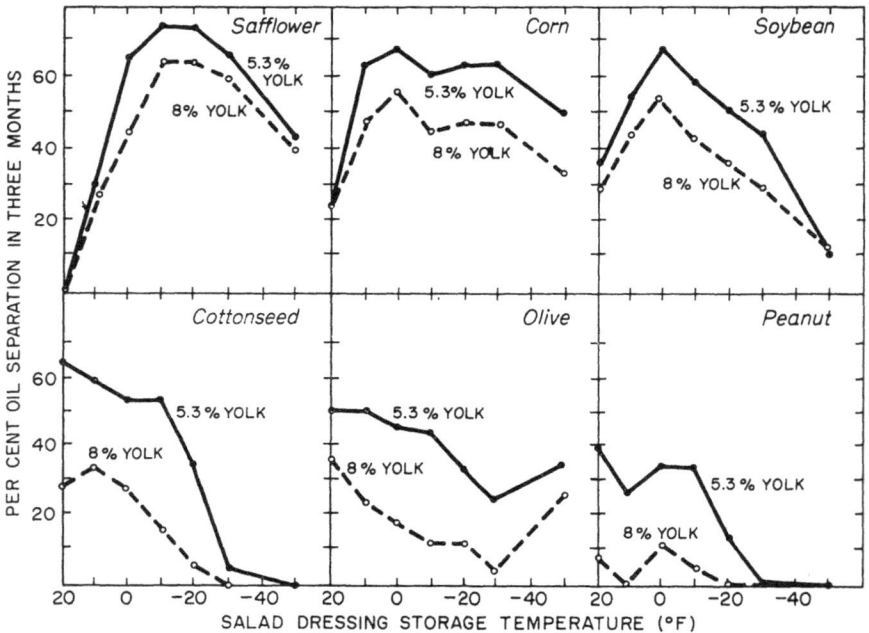

Courtesy H. Hanson and U. S. Dept. Agr.

FIG. 26.   EFFECT OF TYPE OF OIL AND AMOUNT OF EGG YOLK ON OIL SEPARATION IN FROZEN AND THAWED SALAD DRESSING

the sauces and gravies is evidently a function of the starch-bearing ingredient, which is usually flour. Waxy cereal starches are very effective as stabilizers, according to Hanson et al. (1951). Waxy rice flour was found to be particularly valuable, giving good initial texture, excellent resistance to freezing, and the best storage life. These workers also found that some non-starchy ingredients were effective as stabilizers. Among these were some seaweed extractives, pectin, and gelatin. In general, however, these were considerably less effective than the waxy cereal starches. The type of oil and the amount of egg yolk also influence separation, as shown in Fig. 26.

Tressler and Evers (1957) did not find any frozen starch puddings or soft custards on the market in 1956, with the exception of the fillings in some pies. Although the existence of good quality "instant" powders for the easy preparation of these foods may have contributed to the situation they observed, Tressler and Evers felt that one of the chief reasons puddings are not frozen is that they tend to separate and curdle when frozen and thawed. The problem is evidently a difficult one to solve. It is partly due to the separation of water as large crystals which do not redistribute upon thawing, and partly due to denaturation of any egg proteins which may be present.

Hanson et al. (1953) were not successful in formulating baked custard-type desserts which would retain all their desirable texture qualities during freezing and thawing. However, they did prepare satisfactory desserts of the cornstarch pudding type by substituting waxy rice flour or gelatin for the usual cornstarch. They gave a series of formulas for these products, which can be adapted to use as fillings for chocolate cream pies and similar dessert items.

CHANGES OCCURRING DURING FROZEN STORAGE

General

The ice crystals initially formed in frozen foods tend to increase in size upon storage. Tressler and Evers (1957) state that a little moisture remains unfrozen in any frozen fruit even at the lowest storage temperature of any commercial warehouse, that is, $-30°F$. They also indicate that crystal growth is accelerated by fluctuating temperatures. However, it can be shown from thermodynamics that crystal growth will occur in the absence of any free liquid and at a constant temperature. The larger crystals tend to become larger at the expense of the smaller crystals through the accretion of molecules passing through the vapor state. This is analogous to the growth of the larger crystals in a saturated solution at the expense of the smaller crystals.

During storage, the molecular water and the bound water in the food tend to sublime, thereby dehydrating the product. Water liberated in this manner either escapes into the atmosphere or adds on to the growing crystals, depending upon the efficiency of the packaging material. Colloidal components of the protoplasm, and particularly the proteins, can be irreversibly precipitated by the dehydration occurring in this manner. Furthermore, removal of part of the cellular water causes the osmotically active solutes to become concentrated to an extent sufficient to damage some of the larger cell components when thawing occurs. Additionally, the gross changes known as "freezerburn" are frequently observed. This term encompasses the major visual and accompanying textural changes resulting from extensive dehydration of the surface tissues. Freezerburn is generally recognized by a change in color and a shrinking of the tissues. Upon consumption, the portion exhibiting freezerburn is often found to be coarser and tougher than normal.

Muscle Tissue.—Freezerburn is an important cause of texture deterioration in frozen meat, fish, and poultry during storage. Desiccation of the surface of frozen animal foods may first produce a heightening of the color but on continued storage a grayish-yellow appearance develops. Along with the changes in appearance, the tissue assumes a spongy texture. These changes are generally irreversible, although in the earliest stages some recovery can be obtained by placing the product in an atmosphere of high relative humidity (Kaess 1961 and Kaess and Weidemann 1961). The texture and appearance changes are both due to the same cause, i.e., disappearance of ice crystals through sublimation leaving voids in the cells and intercellular spaces. These voids refract the light and give rise to noticeable texture discontinuities.

Poultry has been found to be particularly susceptible to freezerburn damage because of the difficulty in obtaining a package which will closely conform to the irregular surface of the product. This problem is obviously accentuated with small, irregularly contoured items such as shrimp. Use of frozen water or brine glazes completely enveloping the individual pieces or blocks of shrimp has been of great help in extending their storage life.

The desirable texture characteristics are retained in properly packaged meat, fish, and poultry for quite long periods of time at freezer temperatures. In fact, the limiting factors are flavor and appearance rather than texture.

Fruit and Vegetable Products

In frozen fruit juices and in fruits packed in sugar, the usual growth of ice crystals occurs during storage, and, in time, a syrupy material sometimes called the metacryotic liquid separates. This viscous fluid appears sooner in slowly frozen juices and fruits in syrup than it does in rapidly frozen juices, sometimes being evident immediately after freezing in the former products. According to Tressler and Evers (1957), the composition of this liquid in orange juice is fairly constant at eight per cent acid and 64 per cent carbohydrates. It also can contain enough pectin to form a jelly-like substance which may not liquefy when the juice is thawed. The separation of this material can materially influence the viscosity or "texture" of the fluid in syrup canned fruits or of reconstituted juice.

Although enzyme activity is certainly greatly retarded by the temperatures reached in commercial freezer storage, the possibility of texture damage as a result of slow enzyme activity progressing over a long period of time still exists. In frozen pulpy juices such as tomato and orange this may be the cause of the loss of cloud and deposit of sediment which is observed after rather long storage. Bissett *et al.* (1953) found that it was necessary to heat orange juice and orange juice concentrates to at least 190°F. in order to obtain complete stabilization of cloud. They related the protective effect of this temperature to destruction of pectinesterase. Atkins *et al.* (1953) observed a similar phenomenon in orange and grapefruit juices. It would appear that the viscosity of the juice when reconstituted and thawed is also dependent upon the activity while frozen of the enzymes acting upon pectic substances.

Fruit and vegetable foods are, as a rule, less subject to freezerburn than are meat, poultry, and fish. However, desiccation sufficient to damage texture quality can occur in extreme cases.

Cereal Products and Confections

Chemically leavened doughs tend to lose much of their available carbon dioxide upon storage. This is simply a result of the diffusion of the gas into the surrounding atmosphere because of the much greater carbon dioxide tension in the dough. The effect can be considerably retarded by using packaging material which is impervious to carbon dioxide and by reducing the head space to the minimum feasible amount. Use of very slowly reacting acidic components in the leavening system helps to reduce carbon dioxide loss by permitting the establishment of an alkaline environment in the dough. Loss of carbon dioxide leads to a denser baked product having poorer texture.

Yeast-leavened doughs undergo a progressive loss of fermentation capacity during frozen storage. In addition to the usual causes of cell damage which have been discussed previously, translocation of moisture vapor through the inoperative semipermeable membrane of the yeast leads to progressive disturbances of metabolic patterns dependent upon osmotic control. The injured and killed yeast cells emit compounds such as reduced glutathione which may have a deleterious effect upon the gluten network manifested by rupture of the vesicle walls when the product is baked.

In the study by Woodroof which was previously quoted, it was found that all candies with the exception of hard candies tended to dry out during extended frozen storage unless they were sealed in moisture-proof packaging materials.

CHANGES OCCURRING UPON THAWING

Upon thawing, most muscle tissue tends to lose a certain amount of fluid as drip. The amount of drip becomes greater as the storage period increases unless desiccation has been excessive. The tendency to leak, or form drip, varies with the amount of cut surface—whole fish and poultry exhibit less drip than cut pieces. Fish drip more than meat or poultry and coarse fish leak somewhat more than those of finer texture, according to Tressler and Evers (1957). Fish with low pH drip more than those of higher pH. Tarr (1947) indicates that very slow thawing increases protein denaturation in frozen fish with a resultant higher drip. Loss of fluid in this manner clearly affects the texture of the muscle tissue by rendering it tougher and coarser.

Most fruits leak to a considerable extent upon thawing, but vegetables do not exhibit this phenomenon to any great degree and starchy vegetables such as peas, corn, and potatoes have no drip. Because of the extensive cellular damage which occurs during freezing and frozen storage, fruits and vegetables are very susceptible to attack and structural breakdown both by internal enzymes, such as those acting upon pectic substances, and by micro-organisms. For this reason, frozen fruits and vegetables should be thawed immediately prior to use (or cooking) in a manner which does not require long standing in a semifrozen condition.

All baked foods should be thawed as rapidly as possible, preferably in a current of warm air. Except in cakes and other rich products, the staling reaction is accelerated at temperatures near freezing and the longer the product remains within this temperature range the more texture will be damaged. Unbaked yeast leavened doughs should also be thawed promptly since the disturbed osmotic relationships which exist in and around the yeast cells are very harmful to their fermentation processes and other life activities.

Confections pose few problems in the thawing process. Woodroof (1955) indicates that sugar-blooming may result if moisture is allowed to condense on them during the thawing period. Sugar bloom affects the texture of the candy by forming atypical gritty areas on the surface.

SUMMARY

Some disagreement on the value of increasing the rate of freezing appears in the literature. Tressler and Evers (1957) summarized the majority view as follows:

(1) A careful examination of frozen fish and meat usually discloses measurable quality differences between rapidly frozen and slowly frozen products which before freezing were of comparable quality. In the case of vegetables, little difference in quality can be noted between the rapidly frozen and slowly frozen product, provided the slow freezing is sufficiently rapid to check the growth of micro-organisms.

(2) During cooking, most of these differences in quality disappear.

(3) The rate of freezing effects greater differences in quality in lean fish, and in thin cuts of beef and veal, than in most other foods.

(4) Moderately rapid freezing produces frozen food of good quality.

It seems to be generally agreed that the lowest practicable storage temperature is the best. For most foods, 0°F. is recommended. Moisture-vapor-transmission resistant packaging material applied in a manner which reduces to a minimum the free air space around the product is desirable in order to retard desiccation and the associated decline in texture quality. Irregularly shaped small pieces which cannot be packed in this manner can be coated with frozen water or brine.

Under the best of circumstances, thawing is a period of deteriorative changes in the texture of most foods. Destructive osmotic effects, chemical and enzymatic changes, and attack by micro-organisms are facilitated. Therefore, the thawing process should be accomplished in the most expeditious manner possible which does not require raising the temperature of the product unduly.

BIBLIOGRAPHY

ALBRECHT, J. J., NELSON, A. I., and STEINBERG, M. P. 1960A. Characteristics of corn starch and starch derivatives as affected by freezing, storage, and thawing. I. Simple systems. Food Technol. *14*, 57–63.

ALBRECHT, J. J., NELSON, A. I., and STEINBERG, M. P. 1960B. Characteristics of corn starch and starch derivatives as affected by freezing, storage, and thawing. II. White sauces. Food Technol. *14*, 64–68.

ALMÁSI, E. 1952. Investigations on the permeability of cells. Élelmezési Ipar *6*, 82–86.

ASSELBERGS, E. A., MOHR, W. P., and KEMP, J. G. 1960. Studies on the application of infrared in food processing. Food Technol. *14*, 449–453.

ATKINS, C. D., ROUSE, A. H., HUGGART, R. L., MOORE, E. L., and WENZEL, F. W. 1953. Gelation and clarification in concentrated citrus juices. III. Effect of heat treatment of Valencia orange and grapefruit juices prior to concentration. Food Technol. *7*, 62–66.

BECHTEL, W. G., and KULP, K. 1960. Freezing, defrosting, and frozen preservation of cake doughnuts and yeast-raised doughnuts. Food Technol. *14*, 391–394.

BISSETT, O. W. 1958. Processing freeze-damaged oranges. Proc. Florida State Hort. Soc. *71*, 254–259.

BISSETT, O. W., VELDHUIS, M. K., and RUSHING, N. B. 1953. Effect of heat treatment temperature on the storage life of Valencia orange concentrates. Food Technol. *7*, 258–260.

BOYLE, J. L. 1959. The stabilization of ice cream and ice lollies. Food Technol. in Australia *11*, 543–551.

BROWN, R. W., HUMBERT, E. S., and GIBSON, D. L. 1959. Ice milk. Development of a formula satisfactory for both soft-serve and hardened product. Can. Dairy and Ice Cream J. *38*, No. 5, 66, 68, 70, 72.

COLE, L. J. N., KEUEPFEL, D., and LUSENA, C. V. 1959. Freezing damage to bovine cream indicated by release of enzymes. Can. J. Biochem. Physiol. *37*, 821–827.

DAVIS, J. G., HANSON, H. L., and LINEWEAVER, H. 1952. Characterization of the effect of freezing on cooked egg white. Food Research *17*, 393–401.

DAWSON, E. H., GILPIN, G. L., and REYNOLDS, H. 1950. Procedures for home freezing of vegetables, fruits, and prepared foods. U. S. Dept. Agr. Handbook 2.

DEATHERAGE, F. E., and HAMM, R. 1960. Influence of freezing and thawing on hydration and charges of the muscle proteins. Food Research *25*, 623–629.

DIETRICH, W. C., OLSON, R. L., NUTTING, M., NEUMANN, H. J., and BOGGS, M. M. 1959. Time-temperature tolerance of frozen foods. XVIII. Effect of blanching conditions on color stability of frozen beans. Food Technol. *13*, 258–261.

HANSON, H. L., CAMPBELL, A. A., and LINEWEAVER, H. 1951. Preparation of stable frozen sauces and gravies. Food Technol. *5*, 432–440.

HANSON, H. L., and FLETCHER, L. R. 1960. Preparation of salad dressings stable to freezing and frozen storage U. S. Dept. Agr., Western Utilization R. and D. Div. Mimeo. Bull.

HANSON, H. L., NISHITA, K. D., and LINEWEAVER, H. 1953. Preparation of stable frozen puddings. Food Technol. *7*, 462–465.

KAESS, G. 1961. Freezerburn as a limiting factor in the storage of animal tissue with livers frozen without weight loss. Food Technol. *15*, 122–128.

KAESS, G., and WEIDEMANN, J. F. 1961. Freezerburn as a limiting factor in the storage of animal tissue. II. Histological study of freezerburn of liver. Food Technol. *15*, 129–133.

LAW, N. H. 1955. Palatability of chilled and frozen beef. Food Manuf. *30*, 187–189.

LEE, F. A., GORTNER, W. A., and WHITCOMBE, J. 1946. Effect of freezing rate on vegetables. Appearance, palatability, and vitamin content of peas and snap beans. Ind. Eng. Chem. *38*, 341–346.

MACARTHUR, M. 1945. Freezing of commercially packaged asparagus, strawberries, and corn. Fruit Products J. *24*, 238–240.

MOORJANI, M. N., MONTGOMERY, W. A., and COOTE, G. G. 1960. Correlation of taste panel gradings with salt-extractable protein of frozen fish fillets. Food Research *25*, 263–269.

PEARCE, J. A., and LAVERS, C. G. 1949. Liquid and frozen egg. V. Viscosity, baking quality, and other measurements of frozen egg products. Can. J. Research *F-27*, 231–240.

PENCE, J. W., and HAMAMOTO, M. 1959. Studies on the freezing and defrosting of cakes. Food Technol. *13*, 99–106.

PENCE, J. W., and STANDRIDGE, N. N. 1958. Effects of storage temperature on firming of cake crumb. Cereal Chem. *35*, 57–65.

PETERS, J. A., and FLAVIN, J. W. 1956. New techniques for freezing and storing North Atlantic lobsters. Commercial Fisheries Review *18*, No. 7, 22–23.

PETERSON, A. C., and GUNDERSON, M. F. 1960. Role of psychrophilic bacteria in frozen food spoilage. Food Technol. *14*, 413–417.

ROUSE, A. H., and ATKINS, C. D. 1955. Pectin esterase and pectin in commercial citrus juices, as determined by methods used at the Citrus Experiment Station. Univ. Florida Agr. Expt. Sta. Bull. *570*.

ROUSE, A. H., ATKINS, C. D., and MOORE, E. L. 1958. Chemical characteristics of citrus juices from freeze-damaged fruit. Proc. Florida State Hort. Soc. *71*, 216–219.

SAWANT, P. L., and MAGAR, N. G. 1961. Studies on frozen fish. I. Denaturation of proteins. J. Food Sci. *26*, 253–257.

SISTRUNK, W. A., CAIN, R. F., VAUGHAN, E. K., and LAGERSTEDT, H. B. 1960. Factors contributing to the breakdown of sliced frozen strawberries. Food Technol. *14*, 640–643.

TARR, H. L. A. 1947. Preservation of quality of edible fish products. Fisheries Research Board Can., Pacific Coast Stas., Progr. Repts. *71*, 15–20.

TRESSLER, D. K., BIRDSEYE, C., and MURRAY, W. T. 1932. Tenderness of meat. I. Determination of relative tenderness of chilled and quick-frozen. Ind. Eng. Chem. *24*, 242–245.

TRESSLER, D. K., and EVERS, C. F. 1957. The Freezing Preservation of Foods. Avi Publishing Co., Westport, Conn.

WEST, L. C., TITUS, M. C., and VAN DUYNE, F. O. 1959. Effect of freezer storage and variations in preparation on bacterial count, palatability, and thiamin content of ham loaf, Italian rice, and chicken. Food Technol. *13*, 323–327.

WINTER, J. D., and TRANTANELLA, S. R. 1958. Effect of packaging on keeping quality of frozen bread and cake. Package Eng. *3*, No. 8, 32–36.

WOODROOF, J. G. 1938. Microscopic studies of frozen fruit and vegetables. Georgia Expt. Sta. Bull. *201*.

WOODROOF, J. G. 1955. Freezing candies. Georgia Expt. Sta. J. Series *278*.

WOOLRICH, W. R., and BARTLETT, L. H. 1942. Quick and flash freezing of foods. Mech. Eng. *64*, 647–653.

CHAPTER 13

Dehydration

INTRODUCTION

Dehydration of foodstuffs has been practiced for centuries, but only in the last two or three decades have significant advances in the technology been made. Until recently, artificially dehydrated foods were used only for very special purposes, especially for the military, where economic considerations were not of paramount importance, but now an ever-increasing number of specialty foods are being offered through regular commercial channels.

The advantages of dehydration in preserving texture have been of minor influence in stimulating its use as compared to the other improvements and advantages resulting from this processing technique. In fact, dehydration rarely has shown better results than the more conventional preservation methods in promoting desirable texture in the ready-to-eat food. The superior qualities of dehydrated foods are more likely to be their flavor or appearance, or their merchandising characteristics (e.g., improved storage stability, less stringent packaging requirements, and less weight to transport).

This chapter contains discussions of field-dried foods such as cereal grains as well as those which are "artificially" dehydrated. It has been found desirable to include some background material on dehydration methods in order to make the discussion of texture changes more meaningful.

PRE-DRYING TREATMENTS

Either blanching or cooking is commonly resorted to in order to destroy enzymes in foods intended to be dehydrated. The effects of these treatments on the texture of the food are described in detail in Chapter 11. In general, when blanching or cooking is combined with dehydration, the effects of the two processes are additive. The advantages of low-temperature dehydration are mostly lost when cooked products are used as the raw materials. The improved porosity of freeze-dried substances still exists under these conditions, of course.

Freezing is used before dehydration in the case of freeze-drying and afterwards in the case of dehydrofrozen products. If applied to partially dried substances, the effect of freezing on texture can be expected to be greatly reduced. The size of the ice crystals will be reduced and their tissue-distorting effects will thus be minimized. However, the marked texture deterioration which often accompanies the freezing of some high-moisture materials will be just as significant in material later dehydrated. These effects are discussed fully in the preceding chapter. Some information exists to show that relatively high temperature freezing of meats prior to freeze-drying yields a more satisfactory freeze-dried product than does low-temperature freezing. This is said to be true because rehydration of the frozen-dried product is facilitated by the relatively large voids resulting from the formation of large ice crystals. Apparently little tissue damage results from large crystal formation in the case of meats which are to be subsequently dried (Doty 1961).

Fruits scheduled to be dehydrated are often exposed to an atmosphere high in sulfur dioxide (i.e., "sulfured") in order to reduce the enzymatic darkening which otherwise occurs during drying and to secure certain other advantages. The effects of this procedure on textural qualities have apparently not been examined in any detail. It is known that sulfuring destroys the semipermeability of the cell membrane with consequent effusion of cell fluids into the interstitial channels and spaces. The loss of cell turgor is irreversible.

Whole fruits are sometimes immersed for short periods (less than a minute) in a hot sodium hydroxide solution to break open the skin and thus provide pathways through which vapor can leave the fruit during drying. The rate of dehydration can be greatly increased by application of this technique. The surface layers of the fruit are cooked and the skin is fragmented and tenderized. Pectin is demethylated in a layer of flesh whose thickness depends upon the intensity of the treatment. Proteins are denatured and starches gelatinized. The interior of large fruits, such as prunes, may remain completely unaffected by the lye dip with a resultant textural discontinuity between the surface and the region around the seed. Much of this non-uniformity is obscured by the effects of the treatments applied subsequently so that it is not apparent in the finished ' product.

Since this is not a treatise on dehydration procedures in general, a discussion of the merely mechanical pre-treatments of pitting, slicing, peeling, etc. is not pertinent.

TYPES OF RAW MATERIALS

For the purposes of this section, dehydrated foods will be divided into the categories of non-cellular liquids, pulps containing whole cells, and large pieces.

Non-cellular liquids, including milk, beef extracts, gelatin, some fruit juices, and the like, respond in a rather uniform manner to dehydration. They tend to form glassy or pulverulent sheets or films. Exceptions are observed when viscous liquids are dehydrated under reduced pressure and when these fluids are freeze-dried. In either case, a spongy, aerated mass is formed. The preferred dehydration technique is spray-drying. Rehydration of non-cellular liquids dried at high temperatures is often difficult, since the powders tend to agglomerate on first contact with water, yielding a gritty, non-uniform suspension. This textural defect can sometimes be avoided by including a diluent of slowly soluble crystals.

Tomato purées, citrus juices, potato slurries, and similar raw materials comprise the class of "pulps containing whole cells." From the textural standpoint, these products are distinct because the whole cells contribute special viscosity characteristics to the fluid. It is possible to dehydrate these materials in a manner which will not destroy the semipermeable membranes of the cells (although drying will always make them at least temporarily non-functional). Proper rehydration will then restore the textural characteristics dependent upon cell turgor. Many of the other problems encountered in dehydrating and rehydrating these products are very similar to those observed with non-cellular liquids.

The category of "large pieces" includes whole fruits (raisins, prunes), whole vegetables (peas, beans), cut pieces of fruits and vegetables, meat pieces (hamburger size and larger), shellfish (shrimp), fish pieces, etc. The principal problems affecting texture involve the maintenance of a uniform consistency throughout the piece, avoiding case-hardening on drying, and dry-cores after rehydration.

SEQUENCE OF EVENTS IN DEHYDRATION

Thermal Drying Methods

Freeze-Drying Methods.—Many advantages accrue from removal of water vapor while the product is frozen. The many deleterious changes due to high temperatures are, of course, entirely avoided. Proteins are not denatured and so recover most or all of their original moisture content during rehydration. Semipermeable membranes can, under favorable conditions, become re-established with consequent recovery of cell turgor. Gel structures retain their undried shape and size, and their capacity for absorbing water. Lipids are not melted and redistributed. As a consequence of all of these factors, the piece has substantially the same dimensions when dried as it had in the moist state, and penetration of water into the porous mass is greatly facilitated. If the material to be dried is in fluid form, and if the water content is sufficiently high, a mass having a large proportion of voids is created, and the rehydratability is thereby improved.

The drying process does not reverse the damage done by freezing, and rapid temperature reduction during the latter process is desirable if tissue destruction caused by the formation of ice crystals and the separation of solutions having high osmolalities is to be minimized. In addition, the formation of small crystals increases the rate of sublimation during subsequent processing. Large pieces of meat are frequently pierced by spikes attached to the heating plates in order to improve transfer of thermal energy and to increase the number of paths for the exit of vapor, but the holes which are formed usually close when the pieces are reconstituted and there is little or no observable effect on texture.

In the initial phase of drying, the surface moisture and the water in the various kinds of tubules are removed. Since the natural pathways of fluid flow are not shrunken, as they may be in foods dried at higher temperatures, an efficient route of penetration for water is available. Next, the moisture in the cell walls and the underlying semipermeable membranes sublimes. It can reasonably be expected that the membranes will retain the potentiality of selective transport so that, when properly rehydrated, they will bar the leaching from the cell of osmotically active substances. Because of this function, it may be possible to secure cell turgor pressures approaching those in the fresh cells with resultant desirable effects on texture.

The intermediate stage of freeze drying can be considered to include the removal of water vapor from the crystals of ice and eutectic solutions within the cell itself. These crystals leave microscopic voids in the cytoplasm while higher temperature drying tends to cause shrinkage of the gel as water is removed. Probably most gel formations remain capable of re-establishment in approximately the original form although full attainment of this capability in practice may be difficult. Doubtless, oil droplets which are normally separated by a continuous aqueous phase will tend to coalesce on warming after the continuous phase is inactivated by the removal of water, just as they do in foods dehydrated by other methods. Texture characteristics can be expected to change as one result of this effect. Crystallinity of polymeric molecules increases because of the freeing of loci normally hydrogen-bonded to water molecules (i.e., solvated). These loci are thus made available for bonding with similar sites on adjacent polymer molecules so that micelle formation is enhanced. Crystalline regions developed in this way (or in any manner) will not rehydrate, and texture deterioration results.

If drying is continued under sufficiently rigorous conditions, the monomolecular layer of water molecules which adheres tenaciously to the surface of every hydrophilic substance will be removed. Considerably more energy is required for this process than for removing crystalline water. Poorly understood effects on texture occur as a result of extreme dehydration. The rate of non-enzymatic browning and the development of oxidative rancidity are accelerated, but such reactions affect flavor and appearance more than they do texture. There is evidence that abrupt increases in friability, equivalent to a decrease in the strength of bonding between particles, occurs at very low moisture contents. Stresses which may be released by cracking or "checking" may also be set up.

Drying at Intermediate Temperatures

Techniques requiring the removal of water from products whose temperature never rises above about 140°F. can be regarded as intermediate temperature methods. The distinction is important from a texture standpoint because proteins can be kept undenatured at these temperatures and it is possible for the cell membrane to retain a potential capacity for selective transport of ionic substances. However, shrinkage of macrostructures and other tissue changes distinguish such products from those prepared by freeze drying. Among the procedures falling in the intermediate temperature class are sundrying of fish, fruit, and the like, and some vacuum dehydration methods. Not many liquids are dried by these techniques.

Intermediate temperature processes are superior to freeze dehydration at least in the fact that tissue damage by ice crystals does not enter into the picture. This can be a considerable advantage in ripe fruits where the cell walls are likely to be very weak and susceptible to mechanical damage. On the other hand, enzymatic processes and the growth of micro-organisms can continue unabated during the drying period. If the product is cooked, blanched, or sulfured, some of the possible advantages of using mild temperatures are lost. As a practical matter, most of the foods dried at such temperatures are subjected to one or more of these pre-drying treatments.

In the initial stages of the dehydration process, if a relatively large piece is being dried, free water is removed from the surface of the piece and from the network of transport vessels. The loss of water is accompanied by shrinkage and collapse of the vessels in many instances, interrupting potential entry routes for the rehydrating fluid. Vapor passes through intact cuticle with extreme slowness, so the initial loss of moisture occurs through the cut surfaces of fruit, or through breaks in the skin. This differential rate of initial water loss can result in texture differences between the flesh at the cut surface and that next to the skin, even after complete equilibration has occurred, because the enzymatic reactions and other changes affecting textural structures will tend to occur at varying rates in these regions.

If a product is brought to a low enough temperature, all of the solutes will remain fixed during freeze dehydration. Damage of proteins and other gels due to unfavorable concentrations of osmotically active materials developing as the water is progressively removed is thus prevented, although it is true that transient high concentrations may occur during rehydration. The protective effect of solute fixation is not observed in intermediate temperature drying. On the contrary, the aqueous phase is gradually concentrated and may remain in contact with labile gels for relatively long periods. Thus conditions favorable to damage of textural substances exist during the middle drying stage.

Another consequence of the removal of solvent is the enhancement of enzymatic activity and other reactions. Enzymes attacking pectin, proteins, starches, cell-wall structural elements, etc. may affect texture of the food under these conditions. However, permeability barriers between enzyme and substrate may prevent the full effect of the concentration phenomena from being exerted. Reduction of the water content also facilitates crystallization or micelle formation by polymeric molecules as intervening hydrogen-bonded water is removed and the polymers are brought into contiguity.

Evidently it is possible under favorable conditions to complete the middle drying stage without irreversibly inactivating the semipermeable membranes. Such an effect is observed in the drying of active dry bakers' yeast, in which the moisture has been brought to about 7 or 8 per cent by applying temperatures near 110°F. The semipermeable membranes are inactive at this moisture level but the selective transport function can be rapidly re-established by proper rehydration. Similar results can probably be obtained in drying the vegetative cells of other microorganisms. It would appear to be at least theoretically possible to prepare a dried multicellular food which would be capable of attaining its original turgor upon rehydration. There are no published data which permit us to judge the practicality of the scheme.

Shrinkage of gels and micro-structures proceeds as the drying process continues. Although these substances may be chemically unchanged and capable of absorbing their normal complement of solvent, the shrinkage makes it much more difficult for water to penetrate throughout the mass, and prolonged soaking or high temperature treatment may be necessary before the product achieves an acceptably swollen condition and soft texture.

Most foods cannot be brought to very low moisture contents by the methods discussed in this section. At a few per cent moisture content, the equilibrium relative humidity becomes so low that it is impossible to achieve dehydrating atmospheres by gentle heating of the ambient air (or product). If it is desired to remove the monomolecular layer of water, more rigorous methods are required. These might involve higher temperatures, chemical or refrigeration drying of the air, or high vacuum treatment.

Drying at High Temperatures.—Drying methods during which the product temperature exceeds about 140°F. at some time can be considered high-temperature techniques. Many commercially important processes fall into this category. Drum-drying of infant cereals, spray-drying of milk, hot air-drying of fruit pieces, and belt-drying processes for potato granules are examples of high-temperature procedures. It would be expected that texture changes (as well as changes in flavor and appearance) would be accentuated by increases in the drying temperature, and the principal reasons for choosing these methods are economic ones. When cooked foods are to be dehydrated, it is possible to minimize this high-temperature deterioration by taking the raw material to the drying equipment in a fresh or partly-cooked state.

The time required for drying is determined not only by the temperature of the raw material (usually close to the boiling point of water) but also by the rate of air flow over the substance (if air is the heat transfer medium), and the maximum length of travel of water molecules in the piece (i.e., the piece dimensions). The rapidity with which water vapor is removed from the hot material is of great importance in limiting the amount of texture deterioration that will occur. In all of these processes, the food is at a temperature conducive to degeneration and the water must be removed rapidly if acceptable texture is to be retained. On the other hand, exposure to drying temperatures after nearly all of the water has been removed is particularly harmful. Drum-drying, and especially spray-drying make use of relatively short exposure times and it is these processes which have been of most value commercially.

The principal features of high-temperature drying from the texture point of view are the extensive degeneration reactions occurring during processing. Denaturation of proteins, gelatinization of starch, melting of lipids, browning reactions, and other degradative processes are predominant. Collapse and shrinkage of cells and macro-structures are usual. Rehydration of the dried material is hampered by these changes, although, with fluid foods, the problem can be partially alleviated by grinding the dried material to an intermediate particle size. Hydrophilic properties of most gels are reduced by these treatments.

Drying conditions in this group of techniques preclude any retention of semipermeability in the cell membrane. Consequently, cell turgor pressure cannot play a role in texture of foods dehydrated by high temperature methods although hydrophilic colloids bound to internal structures in the cell or polymers too large to diffuse through the voids in the cell wall can still cause some distention of the cell. Destruction of the selective transport function while most of the original moisture is present in the cell allows leaching of soluble materials. These solutes may then be dried on the surface of the cell or in interstices, or they may be translocated within the cell with deleterious effects on osmotically sensitive structures.

The type of processing exemplified by the frying of potato chips is clearly a form of dehydration although it is more common to regard it only as a cooking technique. The process is speeded by using oil as the heat transfer medium. The decrease in speed of water vapor removal due to surrounding the chip with oil is obviously not important. Changes occurring in the product are, in many ways, analogous to those in drumdrying.

Non-thermal Drying Methods

All of the dehydration methods described in the preceding section involved the application of heat to the product even though the product might remain frozen throughout the process. The methods to be discussed in the present section do not require the deliberate addition of thermal energy to the system. Naturally dry (field-dried) products such as cereal grains lose moisture principally as the result of physiological processes associated with maturation, although evaporation also plays a part. This large and economically important class of dried products is not of much practical interest so far as texture changes are concerned since they are unavailable in non-dried form (sweet corn being an exception) and are almost always used in moist-cooked form.

Some foods can be dried by exposing them at room-temperature or below to atmospheres with very low relative humidities. Desiccation over calcium chloride is an example. So far as dehydration conditions are concerned, this treatment is very mild and minimal texture deterioration occurs. For practical purposes, however, drying in this way is unsuitable because of the extremely slow rate of moisture loss. In fact, most foodstuffs spoil before they lose enough water to recome resistant to attack by micro-organisms. It might be possible to use such processes in combination with epoxide sterilization, which has little effect on texture, to yield high-quality products, but this has apparently not been tried.

Dehydration by washing with water-miscible solvents has been tried. Ethanol, methanol, and acetone have been used. Each of these solvents removes non-volatile constituents in addition to water and other volatiles, differing in this respect from all of the methods described previously. The function of the semipermeable membrane is invariably destroyed by solvent extraction. Special techniques must be employed to remove the last traces of solvent, frequently causing side results which offset all of the beneficial effects of low temperature treatment. Gels are usually shrunk and collapsed by solvent extraction. In spite of the disadvantages, solvent dehydration is being used to dehydrate fruit "nuggets." It is also applied to whole fish to yield a powdered "protein concentrate." In the latter case, texture of the original product is completely obscured.

LEVELS OF DEHYDRATION

The proportion of the total moisture removed is closely related to texture and most other properties. For the purposes of this chapter, it is useful to divide dehydrated foods into three categories: (1) high moisture products, (2) low moisture products, and (3) dry products.

High Moisture Products

Significant advantages are sometimes secured by stopping the dehydration process at a stage when the food is still in a moist condition. Dehydrofreezing, the process by which fruits and vegetables are dried to 50 per cent or less of their fresh weight and then frozen, preserves the texture of apples better than either dehydration or freezing alone, at least in some cases. In addition these partially dried fruits soften less when blanched. The moisture content of dried fruits such as raisins and apricots may be intentionally raised before packing, either to improve the texture of the product when it is eaten directly from the carton, or to facilitate its use in bakery products and the like. For the former purpose, a moisture content in excess of 20 per cent is usually desirable. Tomato paste is another economically important product which is only partially dehydrated.

These products are generally resistant to texture changes originating in bacterial and yeast activity, a result of the high osmotic pressure of the aqueous phase, but they are susceptible to fungal damage and, subsequently to growth of other microbiological forms as additional water is released by the metabolic activity of the molds. For this reason, it is necessary to heat sterilize, freeze, or apply other preservative methods in addition to dehydration. It is characteristic of partially dehydrated products that their equilibrium relative humidities are fairly high and they can lose moisture to the atmosphere at relative humidities well within the normal range.

Low Moisture Products

Most dehydrated products, as well as naturally dry foodstuffs such as field-dried peas and beans, cereals, and nuts, fall into this group. They are characterized by resistance to mold growth and by stability of moisture content at relative humidities of about 40 to about 70 per cent. The texture of the dried material is either crisp or leathery, depending mostly on the moisture content. Many products tend to be crisp (or hard, if the pieces are large) at moisture contents below about five per cent.

There is enough water in all low moisture foods to form at least a monomolecular layer, and enzymatic activity can often proceed, although at a greatly reduced rate. Texture qualities may be altered by these processes, as described elsewhere in this chapter.

Dry Materials

When the moisture content of a substance is reduced to a level insufficient to allow a monomolecular layer, special properties result. The product becomes very hygroscopic and is not in equilibrium with any normal atmospheric relative humidity. The texture is altered, resulting in a friable condition in many cases. Oxidation of fats is accelerated. Enzymatic phenomena are completely inhibited. The increase in crystallinity of polymeric compounds may be retarded. The metabolism of microorganisms is, of course, completely halted.

This category embraces many freeze-dried products. In addition, high-temperature dehydrated products such as cookies, crackers, pretzels, and some other bakery products are virtually moisture-free as they come from the oven, but water vapor absorption is so rapid during cooling that they usually reach a moisture content of a few per cent before they are packaged. A similar condition exists with popcorn, potato chips, and certain other foods processed at high temperatures.

CHANGES IN TEXTURE DURING STORAGE

Although a reduction in moisture content to the extent attained in most dehydrated foods slows both enzymatic and non-enzymatic reactions, either as a result of the reduction in ease of diffusion of reactants or due to a reduction in the concentration of the reactant water, these processes are not entirely halted and deteriorative changes resulting in flavor and texture losses can be quite important. Kiermeier and Coduro (1955) showed that amylase was still active in low-moisture foods even at an equilibrium relative humidity of 36 per cent provided that the dry material contained fine capillaries. Water is condensed in these capillaries in amounts sufficient to permit the enzyme to operate. Proteases and pectic enzymes, among other biocatalysts acting on structural elements, could doubtless be effective under similar conditions. Rates of these reactions would be very slow but they could produce detectable effects during the long storage life expected of some dehydrated foods.

Studies of Shumazu and Sterling (1961) on model systems of cellulose and calcium pectate point up the possibility of increasing crystallinity, or ordered molecular aggregation, during storage. During six months of storage at 86°F. samples tended to reach an equilibrium crystallinity value characteristic of the substances. Increases in crystallinity, say of cellulose, starch, or pectins, is correlated with decreases in hydration capacity and therefore, with texture defects.

Mattson *et al.* (1950) related storage changes in the texture of cooked dried peas to the activity of the enzyme phytase. An increase in humidity of the air in the storage chamber caused an improvement in cookability. The higher moisture content of the peas resulting from the increased ambient humidity facilitated the breakdown of phytin by the enzyme. As explained elsewhere, the competition of phytin and pectic substances for divalent ions, particularly calcium, is considered by these workers to have a pronounced effect on the texture of the dried vegetables. A contrary effect in beans was described by Morris and Wood (1956) who found that dry beans having about 13 per cent moisture deteriorated significantly in texture during storage at 77°F. for six months while beans below ten per cent moisture content maintained good texture quality for two years.

Toughening of dehydrated fish during storage has been attributed to the development of cross linkages (hydrogen and disulfide) between denatured protein molecules making up the muscle fibers (Connell 1957). Cross-linking of this sort reduces the swelling capacity of the fiber in much the same way as does the crystallinity discussed by Shumazu and Sterling.

Bhatia (1959) explored the relationship between the non-enzymatic browning reaction and mealiness (a texture defect) in dehydrated cooked pork mince. His results indicated that mealiness is somehow a result of the browning. Evidently the hydration capacity of the proteins is lessened by the degenerative reaction in question. Possibly the non-enzymatic browning reaction plays a more important role than is generally suspected in texture deterioration of low-moisture dehydrated foods. Water is generated by this deteriorative process but it can hardly accumulate in quantities sufficient to alter the texture properties of the foodstuff.

Microbiological activity is probably not present in foods containing less than about ten per cent moisture content, the controlling factors being the amount and type of osmotically active substances relative to the amount of water. Dehydrated foods containing more than this amount of water, such as high-moisture raisins and prunes, may be able to support fungal or yeast activity. Yeasts usually do not have a pronounced effect on texture, though there are exceptions to the rule. Fungi, however, commonly secrete proteolytic, pectinolytic, and amylolytic enzymes in amounts sufficient to cause texture deterioration. These micro-organisms can be combatted by surface sterilization with epoxides in consumer packages of high-moisture dehydrated fruits, a patented process.

REHYDRATION

Although many commercially important dehydrated products such as prunes and dried apricots are normally dehydrated by boiling in an excess of water, this cooking process represents an extreme form of reconstitution procedure which need not be considered in detail. It is more instructive and more generally applicable to follow the sequence of events when rehydration is performed with an excess of room-temperature water.

In the case of fresh whole-cell products dehydrated under gentle conditions, the possibility of re-establishing the selective permeability of cell membranes exists. Even when this does occur, however, it is by no means certain that cell turgor will be the same as in the fresh product. It is reasonable to expect that solutes will remain in the same location relative to one another and to cell compartmentation of whatever sort during dehydration from the frozen stage, but it is not so certain that these normal positions will be occupied throughout the rehydration period even if all of the moisture absorption takes place under the most favorable conditions, i.e., from the vapor state. Solute relationships during the rehydration of active dried bakers' yeast have been studied in more detail than those of any other foodstuff and it is very likely that the situation existing in this food component is representative of that in most gently dehydrated whole-cell foods.

When bakers' yeast is dehydrated under gentle thermal conditions, very few cells are killed (some strains are considerably more susceptible than others). This is a very good indication that the relative positions of the biologically active solutes have been retained during the removal of moisture. If these cells are rehydrated with water in the vapor phase, the total non-viable count (including deaths resulting from both the dehydration and rehydration steps) is frequently less than ten per cent and this figure includes the few per cent of cells which were non-viable in the original suspension. Almost as high a recovery rate is obtained if the cells are rehydrated by dispersing them in water at 110°F., and very little soluble material appears in the extracellular fluid. However, if the cells are dispersed in cold water, say at 35° or 40°F., virtually all of the fermentative capacity is lost—some cells being killed and others reduced to an extremely low vitality. If the cells are centrifuged down from the cold suspension, relatively large quantities of nitrogenous material and carbohydrates are found in the supernatant liquid.

Of course, the influence of yeast on the texture of foods is indirect, and cell turgor of this product plays no role in perceived texture. The importance of the phenomena observed during the rehydration of yeast is in their application to freeze-dehydrated fresh fruit and the like. If the cell membranes are not selectively permeable in the dehydrated stage, the rapidity with which they resume their active state upon contact with excess water will affect the rate at which solutes are leached from the cell, and, consequently, the osmotic pressure of the cell sap, the turgor of the cell, and the texture of the food.

The author was unable to find any publications describing studies of the relationship of the temperature of rehydration water to the texture of those foods in which cell turgor might be expected to play a significant part. In the absence of information to the contrary, it would seem to be desirable to rehydrate these foods with water vapor where possible, and with warm water in other cases.

Unless they are severely collapsed, dehydrated multicellular foods will contain air-filled spaces of considerable volume resulting from removal of water droplets, shrinkage of tissues, etc. In freeze-dehydrated chunks, these voids constitute a high proportion of the total volume of the food. Water must displace this air in order to contact the hydrophilic colloids which are the principal contributors to texture in most foods. The egress of gases may be blocked by the premature formation of a continuous film around the outside of a piece or of an agglomeration of small particles. This is the process which interferes with ready dispersibility of non-fat dry milk powder. In milk it can be counteracted by deliberately forming large agglomerates of particles which separate the more cohesive components during the early stages of rehydration. Furthermore, hydration of the colloids may be hindered by a coating of lipid material formed when the fatty components of the food melt and spread over the gel networks after dehydration is nearly complete. Such a problem is of serious degree in reconstituting whole milk powder. Any interference with maximum absorption of water is liable to have serious effects on texture. Rehydration in an evacuated container is an obvious and usually impractical solution to the problem of entrapped gas. A simpler approach is to make the dimensions of the piece such that displacement of the gas is facilitated.

As water penetrates into a cell or into a granule, localized concentrations of the solutes progress from supersaturated to dilute. Any of these solutions may represent an unfavorable environment for the proteins and other colloidal substances which are in the vicinity. In extreme cases, permanent insolubilization may result with undesirable effects on texture. Rehydration by vapor, at least in the initial stages, allows the hydrophilic colloids to fully solvate before the ionic substances and the smaller molecules dissolve.

SUMMARY

The changes in texture due to dehydration processes can be classified as (1) thermal effects, and (2) results of moisture loss. If the raw material is maintained at a temperature above freezing but below about 140°F. during the early stages of drying, enzymatic reactions can proceed for a relatively long period, with consequent ill effects on pectic materials, proteins, etc. High temperatures may initially accelerate these changes but the enzymes are soon inactivated and non-enzymatic reactions then predominate.

During the initial stages of dehydration, cell turgor is lost. If the dehydration conditions are sufficiently gentle, and rehydration is also mild, turgor pressure can be re-established. However, during the inactive interval, that is, during the late drying and early rehydration period, solutes can be leached from the cell, or translocated within the cell to yield regions of high concentration.

Denaturation of proteins occurs not only from thermal causes but also as a result of excessive concentrations of solutes which interfere with essential solvation loci and hydrogen-bonding. Extensibility and rehydratability of proteins are permanently altered by denaturation and, as a gross result, the structure containing them tends to become condensed and tougher. Many other colloids behave in a similar manner.

Collapse of macro-structures occurs except in the case of some freeze dehydrated products. Entrance of water into the product is then restricted, and the degree of collapse is a measure of the ease with which the substance can be rehydrated. Freeze dehydration introduces an additional deteriorative step, since distortion and rupture of cell walls and other cell elements by ice crystals is not a feature of other dehydration methods.

During storage the degree of crystallinity of polymeric molecules tends to increase, with loss of the ability to solvate. Browning reactions may occur, causing proteins to lose some of their hydration capacity. Enzyme reactions may occur at a low rate, if at least a monomolecular layer of water is present.

BIBLIOGRAPHY

BAKER, G. L., KULP, J. F., and MILLER, R. A. 1952. Role of pectin as related to dehydration and rehydration of simulated fruit preparations. Food Research 17, 26–45.
BAKER, G. L., and MURRAY, W. G. 1947. Pectinic acids as related to texture and quality of dehydrated food products. Food Research 12, 129–132.

BHATIA, B. S. 1959. Role of products of non-enzymatic browning on development of mealiness in dehydrated cooked pork mince during storage in air. Food Sci. Mysore 8, 309–312.

BOGGS, M. M., and TALBURT, W. F. 1952. Comparison of frozen and dehydro-frozen peas with fresh and stored pod peas. Food Technol. 6, 438–442.

BRADLEY, W. B., and THOMPSON, J. B. 1950. The effect of crust on changes in crumbliness and compressibility of bread crumb during staling. Cereal Chem. 27, 331–335.

BROOKS, J. 1958. The structure of the animal tissues during dehydration. In Fundamental Aspects of the Dehydration of Foodstuffs. MacMillan Co., New York.

CALDWELL, J. S., and CULPEPPER, C. W. 1943A. Snap bean varieties suitable for dehydration. Food Packer 24, 363–368.

CALDWELL, J. S., and CULPEPPER, C. W. 1943B. Dehydrating snap beans. Food Packer 24, 420, 422, 424.

CALDWELL, J. S., CULPEPPER, C. W., HUTCHINS, M. C., EZELL, B. D., and WILCOX, M. S. 1944. Further studies of varietal suitability for dehydration in snap beans. Canner 99, No. 10, 12–15, 30.

COLE, L. J. N., and SMITHIES, W. R. 1960. Methods of evaluating freeze-dried beef. Food Research 25, 363–371.

CONNELL, J. J. 1957. Some aspects of the texture of dehydrated fish. J. Sci. Food Agr. 8, 526–537.

COOLEY, A. M., SEVERSON, D. E., PEIGHTAL, D. E., and WAGNER, J. R. 1954. Studies on dehydrated potato granules. Food Technol. 8, 263–269.

CRAFTS, A. S. 1944. Cellular changes in certain fruits and vegetables during blanch-ing and dehydration. Food Research 9, 442–452.

CRANK, J. 1958. Some mathematical diffusion studies relevant to dehydration. In Fundamental Aspects of the Dehydration of Foodstuffs. MacMillan Co., New York.

DOTY, D. M. 1961. Private communication.

EDE, A. J. 1958. Some physical data concerning the drying on potato strips. In Fundamental Aspects of the Dehydration of Foodstuffs. MacMillan Co., New York.

FLOSDORF, E. W. 1949. Freeze Drying. Reinhold Publishing Corp., New York.

GANE, R., and WAGER, H. G. 1958. Plant structure and dehydration. In Funda-mental Aspects of the Dehydration of Foodstuffs. MacMillan Co., New York.

GOODING, E. G. B. 1960. Dehydration of carrots. Food Manufacture 35, 249–254.

GÖRLING, P. 1958. Physical phenomena during the dehydration of foodstuffs. In Fundamental Aspects of the Dehydration of Foodstuffs. MacMillan Co., New York.

GÖRLING, P., and BEUSCHEL, H. 1959. Shrinkage stresses in drying of gel-like and pasty materials. Chem. Ingr. Tech. 31, 393–398.

HALL, R. C., and FRYER, H. C. 1953. Consistency evaluation of dehydrated potato granules and directions for microscopic rupture count procedure. Food Technol. 7, 373–377.

HAMDY, M. K., CAHILL, V. R., and DEATHERAGE, F. E. 1959. Some observations on the modification of freeze-dehydrated meat. Food Research 24, 79–90.

HARRINGTON, W. O., OLSEN, R. L., and McCREADY, R. M. 1951. Quick-cooking dehydrated potatoes. Food Technol. 5, 311–313.

HAWKE, J. C. 1957. Consumer trials for the determination of the acceptability of dehydrated beef. J. Sci. Food Agr. 8, 197–205.

HEISLER, E. G., HUNTER, A. S., WOODWARD, C. F., SICILIANO, J., and TREADWAY, R. H. 1953. Laboratory preparation of potato granules by solvent extraction. Food Technol. 7, 299–302.

HOGAN, J. T., and PLANCK, R. W. 1958. Hydration characteristics of rice as in-fluenced by variety and drying method. Cereal Chem. 35, 469–482.

HUNT, S. M. V., and MATHESON, N. A. 1958. The effects of dehydration on acto-myosin in fish and beef muscle. Food Technol. 12, 410–416.

KIERMEIER, F., and CODURO, E. 1955. The effect of moisture content on enzyme reactions in low-moisture foods. Z. Lebensm. Untersuch. u. Forsch. *102*, 7–15.

KRAMERS, H. 1958. Rate-controlling factors in freeze drying. *In* Fundamental Aspects of the Dehydration of Foodstuffs. MacMillan Co., New York.

KUPRIANOFF, J. 1958. "Bound water" in foods. *In* Fundamental Aspects of the Dehydration of Foodstuffs. MacMillan Co., New York.

LAZAR, M. E., CHAPIN, E. O., and SMITH, G. S. 1961. Dehydrofrozen apples: Recent developments in processing methods. Food Technol. *15*, 32–36.

LEA, C. H. 1958. Chemical changes in the preparation and storage of dehydrated foods. *In* Fundamental Aspects of the Dehydration of Foodstuffs. MacMillan Co., New York.

MARK, H. 1940. Intermicellar hole and tube system in fiber structure. J. Phys. Chem. *44*, 764–788.

MATTSON, S. 1946. The cookability of yellow peas. A colloid-chemical and biochemical study. Acta Agr. Suecana *2*, 185–231.

MATTSON, S., ÅKERBERG, E., ERICKSSON, E., KOUTLER-ANDERSSON, E., and VAHTRAS, K. 1950. Factors determining the composition and cookability of peas. Acta Agric. Scand. *1*, 40–61.

MILLER, M. W., and FISHER, C. D. 1961. Production of high-moisture raisins. Food Technol. *15*, 276–279.

MORRIS, H. J., and WOOD, E. R. 1956. Influence of moisture content on keeping quality of dry beans. Food Technol. *10*, 225–229.

MORRIS, T. N. 1947. The Dehydration of Food. D. Van Nostrand Co., New York.

MULLINS, W. R., POTTER, A. L., WOOD, A. O., HARRINGTON, W. O., and OLSON, R. L. 1957. A physical test for consistency of potato granules. Food Technol. *11*, 509–511.

OLSON, R. L., HARRINGTON, W. O., NEEL, G. H., COLE, M. W., and MULLINS, W. R. 1953. Recent advances in potato granule technology. Food Technol. *7*, 177–181.

REEVE, R. M. 1943. Changes in tissue composition in dehydration of certain fleshy root vegetables. Food Research 8, 146–155.

REEVE, R. M., and NEEL, E. M. 1960. Microscopic structure of potato chips. Am. Potato J. *37*, 45–52.

REEVE, R. M., and NOTTER, G. K. 1959. An improved method for counting ruptured cells in dehydrated potato products. Food Technol. *13*, 574–577.

ROLFE, E. J. 1958. The influence of the conditions of dehydration on the quality of vacuum-dried meat. *In* Fundamental Aspects of the Dehydration of Foodstuffs. MacMillan Co., New York.

SALWIN, H. 1959. Defining minimum moisture contents for dehydrated foods. Food Technol. *13*, 594–595.

SALWIN, H., and SLAWSON, V. 1959. Moisture transfer in combinations of dehydrated foods. Food Technol. *13*, 715–718.

SHARP, J. G., and ROLFE, E. J. 1958. Deterioration of dehydrated meat during storage. *In* Fundamental Aspects of the Dehydration of Foodstuffs. MacMillan Co., New York.

SHUMAZU, F., and STERLING, C. 1961. Dehydration in model systems: Cellulose and calcium pectate. J. Food Sci. *26*, 291–296.

SIMON, M., WAGNER, J. R., SILVEIRA, V. G., and HENDEL, C. E. 1953. Influence of piece size on production and quality of dehydrated Irish potatoes. Food Technol. *7*, 423–428.

STITT, F. 1958. Moisture equilibrium and the determination of water content of dehydrated foods. *In* Fundamental Aspects of the Dehydration of Foodstuffs. MacMillan Co., New York.

TISCHER, R. G., JERGER, E. W., KEMPTHORNE, O., CARLIN, A. F., and ZOELLNER, A. J. 1953. Influence of variety on the quality of dehydrated sweet corn. Food Technol. *7*, 223–226.

WANG, H., ANDREWS, F., RASCH, E., DOTY, D. M., and KRAYBILL, H. R. 1953. A histological and histochemical study of beef dehydration. I. Rate of dehydration and structural changes in raw and cooked meat. Food Research 18, 351–359.

WANG, H., AUERBACH, E., BATES, V., ANDREWS, F., DOTY, D. M., and KRAYBILL, H. R. 1954B. A histological and histochemical study of beef dehydration. II. Influence of carcass grade, aging, muscle, and electrolysis pre-treatment. Food Research 19, 154–161.

WANG, H., AUERBACH, E., BATES, V., DOTY, D. M., and KRAYBILL, H. R. 1954A. A histological and histochemical study of beef dehydration. IV. Characteristics of muscle tissues dehydrated by freeze-drying techniques. Food Research 19, 543–556.

CHAPTER 14

Radiation

INTRODUCTION

Food technologists have proposed that ionizing radiation in the form of beta or gamma rays be used for inactivating enzymes, as a substitute for thermal sterilization of foods in hermetically sealed containers, as a pasteurizing agent, as a mutagen for producing more desirable varieties of fruits and vegetables, as a means of destroying insect infestation in foods and raw materials, and for altering growth processes in plant materials. In many cases the desired effects are accompanied by undesirable organoleptic changes such as unpleasant odors and tastes and atypical textures. However, the undesirable effects may not be as pronounced as those resulting when other methods are used to accomplish the same result.

Shea (1958) lists the following minimum requirements for some of the common processing goals: Sprout Inhibition, 10,000 rads; Insect Deinfestation, 50,000 rads; Pasteurization, 500,000 rads; Sterilization, 4,800,000 rads. A rad, which is a measure of the dosage of irradiation, is equivalent to the absorption of 100 ergs of energy per gram of material.

Two major theories, or types of theories, have been advanced to account for the action of ionizing radiations on substances. The target theory emphasizes direct action on the affected molecules. There is a localized release of comparatively large amounts of energy as the result of the interaction of the quanta of radiation and the target molecules. This occurs in, for example, a typical cloud chamber experiment, where each water droplet represents the expenditure of 30 to 100 electron volts of energy. The strongest chemical bonds are formed with energies of a few electron volts.

According to the diffusion theory, the action of ionizing radiation on complex systems is indirect. The radiant energy acts upon water molecules to form highly reactive free radicals. Thus H_2O yields the free radicals H and OH. These radicals then diffuse, about the cell to form other free radicals on organic molecules with resultant destruction of function.

Hutchinson (1961) found that a dose of a few hundred roentgen could destroy the activity of sufficiently dilute enzyme systems as effectively as it could destroy living cells. He regarded these results as clear evidence of radiation produced radicals. A particular molecule regarded as a sphere of radius R will be inactivated if a primary ionization occurs within its volume. Radicals formed in the volume of water [equal to about $4\pi R^2 \rho(1 + \rho/R)$] immediately surrounding the molecule will also cause inactivation. It is clear that radiation induced changes can occur in the molecular species responsible for structure and texture as readily as they can occur in genetic material or enzymes.

The great amount of energy required to secure substantially complete inactivation of enzymes makes ionizing radiation impractical for this purpose. Deleterious changes in texture, as well as in flavor and appearance, occur in nearly all products so treated. Exceptions are liquid foods of low viscosity. Since enzymatic changes are frequently the limiting factors in determining the maximum duration of storage, products preserved by radiation must often be heat-treated, dehydrated, or stored at low temperatures (i.e., frozen) in order to minimize these changes. Pearson *et al.* (1960B) found that proteolysis occurred in raw irradiated roasts stored at 76°F. and appeared to be responsible for the poor acceptability and mushy texture of 32-day samples.

So far as texture deterioration is concerned, radiation and the other preservative treatments often appear to be additive in their effects. Pearson *et al.* (1960A) stated that

Irradiation appears to be a contributing factor to the loss of texture in precooked irradiated meat, but the texture becomes even poorer as storage is prolonged. It has been observed that texture loss is not a serious problem in irradiated *fresh* meats and that consistency is good, even after long periods of storage.

Stadelman and Wise (1961) found that cooking before gamma irradiation of adequately aged poultry exerted only minor effects on the shear values of breast meat. The effects of the processing treatments did not appear to be additive in this case.

With present technology, radiation sterilization requires dosages of about 4.6 megarads in order to secure the same degree of "sterility" achieved in thermal canning processes for non-acid foods. However, according to Mehrlich (1961), it may be possible to reduce the treatment to 3.6 to 3.8 megarads. Undesirable flavor changes usually occur at these treatment levels, but the texture of the irradiated food may remain within acceptable limits. Usually the texture changes are not as extreme as those observed in retorted foods but they may be quite different qualitatively. Gillies (1959) observed that the texture of peas was the organoleptic property least adversely affected by irradiation (4.65 megarads).

Pasteurization, deinfestation, and such special applications as the inhibition of sprouting in stored potatoes require the use of much smaller dosages than those necessary for sterilization and enzyme inactivation.

Consequently, such treatments can often be applied without causing detectable changes in texture.

Most investigators of organoleptic changes in irradiated foods have been concerned more with taste and aroma deterioration than with texture alterations. Detectable changes in flavor seem to show up at lower dosages and are more effective than texture deterioration in lowering acceptability. This is by no means a universal rule and some foods, such as lettuce, lose desirable textural qualities at much lower radiation levels than they acquire off-flavors.

The precise locus of action of ionizing radiation on the substances contributing to the texture of foods is not known with any certainty. It appears that rather extensive depolymerization reactions can be initiated with a resultant weakening of structural elements based on pectic substances, proteins, lignins, and cellulose. The selective permeability characteristics of the cell membranes are usually lost at low dosage levels so that cell turgidity is not a factor in the texture of most irradiated foods.

Investigations of model systems and such simple manufactured foods as jellies have provided the most valuable fundamental information about the processes by which ionizing radiation affects food texture.

EFFECTS IN SIMPLE SYSTEMS

Proteins

Kraybill et al. (1960) said that irradiation of proteins with gamma rays or high speed electrons will, in general, result in denaturation, degradation, polymerization, or molecular rearrangement. These effects are frequently seen as changes in viscosity or gel strength of simple systems such as gelatin solutions. Mateles and Goldblith (1958) indicated that doses up to two megareps had relatively small effects on the bloom strength of gelatin. On the other hand, Bolaffi et al. (1959) found that bloom strength and solution viscosity were reduced when dry gelatin was irradiated at up to six megarads. Low bloom gelatins had greater percentage losses in strength than did high bloom gelatin whereas the reverse relationship was true for viscosity changes. Results obtained with heated controls indicated that the changes were not due to the temperature rise observed during irradiation. These workers suggested that the observed effects were probably due to random splitting of the gelatin polypeptide at the peptide linkages. Several reducing compounds inhibited the changes but nitrogen atmospheres or lowered moisture contents did not. Some of the data of Bolaffi et al. are summarized in Table 22.

TABLE 22

EFFECTS ON GELATIN QUALITY OF IRRADIATING IT IN THE PRESENCE OR ABSENCE OF ADDITIVES[1]

Radiation Dose, M Rads	Gelatin	Gelatin + Pectin	Gelatin + Gum Arabic	Gelatin + Dextrin	Gelatin + Sodium Iso-ascorbate	Gelatin + Cysteine	Gelatin + Fructose
Per Cent Loss in Bloom Strength for Increasing Dosages							
High Bloom Gelatin							
2	12.3	2.7*	7.6	7.0	6.4	6.2	6.9
4	26.0	15.8*	17.6*	20.7	22.3	23.8	...
6	38.2	32.9	32.4*	31.7*	32.0*	39.1	23.1*
Low Bloom Gelatin							
2	16.1	17.7	14.8	27.2	14.5	10.0	17.2
4	25.8	34.2	33.9	39.5	27.2	20.0	30.0
6	45.7	43.0	44.5	44.9	40.5	37.8*	...
Per Cent Loss in Viscosity of High Bloom Gelatin							
2	45.5	36.3	50.7	49.1	30.5	26.6	34.8
4	56.9	55.1	52.7	58.8	46.0	41.0	40.6
6	56.5	61.7	58.0	55.2	55.9	51.1	50.3

[1] Adapted from the paper of Bolaffi *et al.* (1959).
* Significantly different (at the five per cent level) from the unsupplemented gelatin.

Lloyd *et al.* (1957) observed a linear relationship between the radiation dose administered to lyophilized wheat gluten and the viscosity of one per cent solutions made from it, but a non-linear function described the relation of dosage to the viscosity of irradiated one per cent solutions, indicating that separate mechanisms were responsible for the two phenomena. Viscosity losses in the gluten solutions were practically independent of the temperature within the range of 32° to 122°F. X-rays were more effective in reducing the viscosity of gluten solutions at the greater dilutions.

As stated by Lloyd *et al.* their data support the activated solvent (diffusion) theory of radiation action. This theory holds that part (perhaps 90 per cent) of the fraction of energy absorbed by the solvent is transferred to solute molecules by a mechanism involving free radicals. However, they also showed that hydrogen peroxide produced by irradiation of solutions containing dissolved oxygen played no part in the mechanism responsible for X-ray induced reduction in the viscosity of gluten solutions. Sulfhydryl-containing reducing agents gave no protection against these changes.

Carbohydrates

Cellulose and pectin are known to be degraded by irradiation, and the amount of degradation as measured by the percentage fall in viscosity is proportional to the logarithm of gamma ray dosage. Glegg and Kertesz (1956) irradiated samples of cellulose and pectin containing 0.32 and

0.75 per cent moisture, respectively, and found that additional degrada-tive changes occurred during storage of the samples at room temperature. The aftereffect was of the same order as the primary effect and occurred only at the lowest moisture contents.

In a study of gamma-irradiated potato starch and the amylose and amylopectin extracted from it, Kertesz et al. (1959) observed a change in viscosity beginning at some dosage near 60,000 rads. The degradation followed a linear relationship with the logarithm of the dosage.

The effects of ionizing radiation on dry pectin, on pectin solutions in the presence and absence of sugars, and on pectin-acid-sugar jellies were the subjects of an investigation conducted by Kertesz et al. (1956). Criteria of change were the results of jelly tests and viscosity measurements. Pectin of 9.4 per cent moisture content was degraded by electron bom-bardment or gamma-irradiation at a dosage of 50,000 rads. When in solution, pectin was degraded by the lowest dosage of gamma rays, 8,300 rads, but this effect was reduced by the addition of sugar. When the pH and the sugar concentration of the irradiated mixture were in the range suitable for jelly formation, the effect was completely eliminated up to 212,000 rads, as determined by sag tests on the jelly and viscosity measure-ments on pectins extracted from the jellies. These workers concluded that de-esterification is not an important factor in degradation of pectins.

Naik-Kurade et al. (1959) studied the effect of cathode ray and gamma irradiation on the viscosities of organic acid-carbohydrate systems as measured by the Brookfield viscosimeter. They found that a marked reduction in the gel strength of pectin gels accompanied increases in dosage, terminating in complete liquefaction in samples irradiated above 7.32 megarads. Consistent with their observations on model systems, the viscosity of apple sauce was found to decrease with increasing radiation dosages. The viscosities of sweetened samples remained higher than those of the unsweetened samples indicating perhaps a radioprotective action of sugar similar to that observed by Kertesz et al. (1956) in sugar-acid-pectin jellies.

EFFECTS ON COMPLEX FOODSTUFFS

Irradiation at the levels commonly employed for preservation will not destroy the bulk of the enzymes present in a food, and texture-affecting changes such as proteolysis in meats and pectinolysis in fruits and vege-tables will continue practically unabated, eventually rendering the food unacceptable. For example, Licciardello et al. (1959A) showed that the shelf-life of blanched chicken irradiated at three megarads was limited by the poor texture resulting from action of proteolytic enzymes.

Milk

Gamma irradiation of milk results in dosage-related increases in sulf-hydryl and disulfide content, and these are accompanied by, or are the cause of, increases in viscosity (Kraybill *et al.* 1960). See Table 23. Electrophoresis of the casein and whey fractions showed that below

TABLE 23

INFLUENCE OF GAMMA IRRADIATION ON THE VISCOSITY AND PROTEIN SULFHYDRYL CONTENT OF RAW SKIMMED MILK[1]

Irradiation Dosage, Megarads	Viscosity, Centipoises	Protein Sulfhydryl, Mg. $\times 10^{-2}$
0	1.498	4.30
0.465	1.533	4.39
2.79	1.574	4.81
5.58	1.574	7.21
9.30	1.715	11.75

[1] From Kraybill *et al.* (1960).

5.58 megarads, the component proteins are denatured or destroyed, while at 5.58 megarads or higher, an electrophoretically immobile component appears in both fractions. At 9.30 megarads, this electrophoretically immobile fraction is the sole material present.

Concentrated irradiated milk stored at room temperature or higher has a marked tendency to gel (Hoff *et al.* 1960A). In milk of normal concentration, or in concentrated milk stored under refrigeration, gelation is not as apparent. The addition of polyphosphates to concentrated (3 to 1) milk before it is irradiated markedly reduces the tendency to gel when stored at 99°F. Pyrophosphates are not effective (Hoff *et al.* 1960B). Analyses of ultrafiltrates of irradiated milk showed that irradiation caused the release of calcium, magnesium, and phosphate ions from complexing with the casein. The addition of tripolyphosphates to the milk caused increased complexing of calcium, magnesium, and phosphate ions with casein but addition of the tripolyphosphate to milk before it was irradiated did not affect the cation-protein complex although it did cause a release of inorganic orthophosphate and other phosphates from the casein complex.

Addition of sodium diphosphate or sodium citrate enhances the rate of gel formation and causes increases in the gel strength of irradiated milk. The divalent ions magnesium and manganese increase time of gelling and decrease gel strength in low concentrations while calcium apparently has no effect. Normal heat processed concentrated milk shows an opposite response to these ions, citrate and phosphate delaying gelling while calcium has a reinforcing effect. Hoff *et al.* (1960A) have suggested that irradiation causes rupture of pyrophosphate linkages essential to the stability of casein micelles. Storage gelation seems to occur as a result of the disturbance of the electrostatic balance of the micellar system by irradiation.

Meats

Fujimaki *et al.* (1961) investigated the effects of gamma irradiation on the chemical properties of the actin and actomyosin of meats. Muscle tissue was irradiated with 0.4 megarad at 40°F. at three different stages: (1) immediately after slaughter, (2) at maximum rigor, and (3) at "rigor off." The actin and two kinds of actomyosin were isolated and determinations made of the contents of sulfhydryl groups and amino acids, the viscosity, the apyrase activity, and the adenosinetriphosphate sensitivity. The results indicated that actin was relatively insensitive to irradiation as compared with actomyosin, which was considered very sensitive. Depolymerization of actomyosin apparently occurred when the meat was irradiated. The molecules seemed to be most sensitive to irradiation at the rigor-off stage. Although Fujimaki *et al.* (1961) did not speculate on the textural effects of the changes they observed, it is reasonable to assume that depolymerization of actomyosin would cause a softening or tenderizing of the meat.

Batzer *et al.* (1959) found that beef irradiated at 2 and 4 megarads developed a soft texture upon aging, especially when it was aged at higher temperatures. When the meat was irradiated at eight megarads it had a slightly rubbery texture. After three months storage at 60° or 90°F., samples treated with either 4 or 8 megarads were very rubbery. Maintaining the sample at very low temperatures during radiation apparently minimizes the damage. Mehrlich (1961) stated that sterilized raw or cooked beef with essentially no change in texture has been obtained on a laboratory scale by irradiating (4.5 megarads) at —312°F., the temperature of liquid nitrogen. It would be expected that diffusion of H and OH radicals would be greatly restricted at these temperatures and that chain reactions would be inhibited.

Lim *et al.* (1959) observed that beta irradiation at 3 or 6 megareps caused significantly lower juiciness scores and apparent fineness of grind in minced pork. The texture also changed in the direction of greater softness, moistness, waxiness, and looseness.

No texture changes unequivocally related to radiation were found in Pacific cod fillets irradiated at dosage levels up to 1.86 megarads (Miyauchi 1960).

Most of the texture changes observed in irradiated meats are consistent with a mechanism involving depolymerization of many of the long chain molecules which form the predominant structural elements of the muscle fibrils. Since the ruptured molecules are moderately long and many breaks would be required to cause a detectable change in texture, some sort of chain reaction mechanism must be postulated in order to explain the occurrence of texture alterations at the dosage levels which have been used. In order to prevent these texture changes and at the same time permit the desired sterilization reactions to occur, some means of interrupting the chain reactions must be used. The low temperature irradiation mentioned by Mehrlich and other investigators is one possibility. Use of interceptor compounds which can diffuse freely throughout the food but which are not taken up by vegetative cells or spores is another approach which might prove to be worthwhile.

Fruits and Vegetables

Crisp fruits and vegetables which depend for their desirable texture qualities on the retention of cell turgidity are usually softened or wilted by relatively small dosages of ionizing radiation. This phenomenon is doubtless due to a high degree of susceptibility of the semipermeable membrane to inactivation by the free radicals generated by the radiant energy.

Hannan (1955) stated that lettuce irradiated with 930 kilorads of cathode rays showed softening. Boyle *et al.* (1957) irradiated seven varieties of apples and five varieties of carrots with 0.0162 to 2.21 megarads of gamma rays. All dosages caused softening in all varieties of apples and carrots. When firmness was measured as the load in pounds required to crush cylinders of the fruit or vegetable, there was found to be a linear relationship between the percentage change in crushing load and the log of the radiation dosages. Similar results were reported by Gillies *et al.* (1957) who showed that irradiation of apple slices caused softening. The slices receiving 0.5 megareps had a better texture quality than those receiving 1.0 megarep.

Quantitative measurements of the softening caused by irradiation of apples, beets, and carrots were reported by Glegg *et al.* (1956). The changes in firmness were recorded in terms of the load required to crush a cylinder of the fruit or vegetable, a procedure similar to that used later by Boyle *et al.* (1957). Using radiation dosages varying from 4,000 to nearly 4,000,000 rads, it was found that no measurable softening occurred below a certain critical dosage. Above this level pronounced softening occurred which was related to the logarithm of the dosage. Values for the threshold dosages were obtained by graphical interpolation of straight-line plots and were found to be 34,700 rads for Gravenstein apples, 166,000 rads for Chantenay carrots, and 316,000 rads for Detroit Dark Red beets. These results would seem to imply that structures or linkages not responsible for texture effects absorb the released energy up to a given point. When these sites are saturated, the texture-affecting loci are attacked.

Lima beans harvested at the prime canning stage were sealed under vacuum in tin cans and subjected to dosages of radiation ranging from one to one hundred megareps in a study conducted by Salunkhe (1957). The turgidity of the pods decreased under irradiation at dosages above 40 megareps. The pods split at the dorsal and ventral suture and the beans became detached. Microscopic examination revealed that the cells had separated from each other at dosages above two megareps, possibly owing to partial or complete destruction of the contents of the middle lamellae. The starch grains themselves showed no visible effect of irradiation but they were extruded from the cells in numbers proportional to the dosage. The protoplasm remained visually unchanged.

Lück et al. (1960) investigated the effect on carrot powder of electronic irradiation at room temperature over a range of one-tenth to one hundred megarads. Above one megarad the cell structure began to disintegrate and there was loss of swelling power. Crude fiber began to decrease at the high dosage of 100 megarads.

Mullins and Burr (1961) reported that a dose of 250 kilorads accelerated softening of onions, but lower levels of irradiation did not do so. Salunkhe et al. (1959) found that gamma irradiation increased the tenderness of asparagus.

Cereal Foods

Most of the studies of irradiation by cereal technologists have been concerned with the effects of deinfestation dosages on wheat and the flour milled from it. Since semipermeability effects are not important here, it might be expected that detrimental changes would not be observed at levels of, say, 50,000 rads. This appears to be the case. Nicholas et al. (1958) showed that loaf volumes of bread made from irradiated flour were not significantly affected up to dosages of 0.5 megareps. However, flour milled from wheat irradiated at 50,000 reps yielded bread of significantly higher specific volume than flour milled from wheat irradiated at 250,000 reps. These observations may point to a semipermeability change. When the germ of wheat is damaged (mechanically, thermally, enzymatically, etc.), it can release substances which reduce loaf volumes by weakening the gluten structure. It is conceivable that the semipermeable membranes of the embryo cells are inactivated by the radiation dosages in question with a consequent release of detrimental substances into the endosperm.

Webb et al. (1961) investigated the effect of ionizing radiation on cakes and biscuits made with milled irradiated wheat. They said that significant differences in texture were found to result from irradiation, but examination of their data indicates that the changes were small. Some of their data are reproduced in Table 24.

TABLE 24

EFFECT OF IONIZING RADIATION ON CAKES MADE WITH MILLED IRRADIATED WHEAT[1]

	Level of Radiation, Rads				
	Control	23,000	46,000	70,000	93,000
Tenderness (panel score)	4.0	4.0	3.9	3.8	3.9
Texture (panel score)	4.0	3.9	4.0	3.8	3.7
Volume (cc.)	2,790	2,792	2,795	2,801	2.794
Compressibility (mm.)	2.14	2.13	2.15	2.21	2.09

[1] Adapted from Webb *et al.* (1961).

SUMMARY

Ionizing radiations may cause changes in the texture of foods by in-activating the semipermeable membranes of the cells or by initiating depolymerization reactions involving pectins, cellulose, proteins, and other compounds responsible for maintaining the structural character-istics of the cells and cell aggregates. The extent of the latter changes appears to be linearly related to the logarithm of the dosage, at least in many model systems. There is some evidence that loss of turgidity de-pends upon a threshold dosage.

It can be concluded that de-infesting or pasteurizing dosages can be used without causing appreciable deterioration in the texture of many foods. Sterilizing dosages can adversely affect the texture of many foods while an improving tenderizing effect can be observed in other instances. There is some evidence that damage due to sterilizing dosages can be minimized by irradiating at very low temperatures (e.g., $-312°F$.).

At practical levels of treatment, radiation does not appreciably inhibit the enzymes naturally present in fresh foods. Some of these enzymes attack the structural elements of the food, and, over a period of time, create unacceptable texture changes. Auxiliary treatments such as freezing, dehydration, or cooking are necessary to secure adequate tex-ture stability and these treatments can affect texture to an extent which may be additive to the changes caused by irradiation.

BIBLIOGRAPHY

ASSELBERGS, E. A., FERGUSON, W. E., and MACQUEEN, E. F.   1958.   Effects of sodium sorbate and ascorbic acid on attempted gamma radiation pasteurization of apple juice. Food Technol. *12*, 156–158.

BATZER, O. F., SLIWINSKI, R. A., CHANG, L., PIH, K., FOX, J. B., JR., DOTY, D. M., PEARSON, A. M., and SPOONER, M. E.   1959.   Some factors influencing radiation induced chemical changes in raw beef. Food Technol. *13*, 501–508.

BOLAFFI, A., MEZZINO, J. F., LOWRY, J. R., and BALDWIN, R. R. 1959. Effects of ionizing radiation on gelatin and the role of various radioprotective agents. Food Technol. *13*, 624–628.

BOYLE, F. P., KERTESZ, Z. I., and GLEGG, R. E. 1957. Effects of ionizing radiations on plant tissues. II. Softening of different varieties of apples and carrots by gamma rays. Food Research *22*, 89–95.

BROWNELL, L. E., GUSTAFSON, F. G., NEHEMIAS, J. V., ISLEIB, D. R., and HOOKER, W. J. 1957. Storage properties of gamma-irradiated potatoes. Food Technol. *11*, 306–312.

DESROSIER, N. W., and ROSENSTOCK H. M. 1960. Radiation Technology in Food, Agriculture, and Biology. Avi Publishing Co., Westport, Conn.

FUJIMAKI, M., ARAWAKA, N., and OGAWA, G. 1961. Effects of gamma irradiation on the chemical properties of actin and actomyosin of meat. J. Food Sci. *26*, 178–185.

GILLIES, R. A. 1959. Organoleptic evaluation of the combined effects of heat and radiation on canned peas. Food Research *24*, 62–67.

GILLIES, R. A., NELSON, A. I., STEINBERG, M. P., MILNER, R. T., NORTON, H. W., and MORGAN, B. H. 1957. Radiation sterilization of apple slices. Food Technol. *11*, 648–651.

GLEGG, R. E., BOYLE, F. P., TUTTLE, L. W., WILSON, D. E., and KERTESZ, Z. I. 1956. Effects of ionizing radiations on plant tissues. I. Quantitative measurements of the softening of apples, beets, and carrots. Radiation Research *5*, 127–133.

GLEGG, R. E., and KERTESZ, Z. I. 1956. Aftereffect in the degradation of cellulose and pectin by gamma rays. Science *124*, 893.

HANNAN, R. S. 1955. Scientific and technological problems involved in using ionizing radiation for the preservation of food. Great Britain Dept. Sci. Ind. Research Food Invest., Special Rept. *61*.

HOFF, J. E., SUNYACH, J., PROCTOR, B. E., and GOLDBLITH, S. A. 1960A. Radiation preservation of milk. X. Studies on the radiation induced gelation of concentrated milk. Methods and effect of some additives. Food Technol. *14*, 24–26.

HOFF, J. E., SUNYACH, J., PROCTOR, B. E., and GOLDBLITH, S. A. 1960B. Radiation preservation of milk. XI. Studies on the radiation induced gelation of concentrated milk. The effect of polyphosphates. Food Technol. *14*, 27–29.

HOUGH, L. F., and WEAVER, G. M. 1959. Irradiation as an aid in fruit variety improvement. I. Mutations in the peach. J. Heredity *50*, 59–62.

HUTCHINSON, F. 1961. Molecular basis for action of ionizing radiations. Science *134*, 533–538.

KERTESZ, Z. I., MORGAN, B. H., TUTTLE, L. W., and LAVIN, M. 1956. Effect of ionizing radiations on pectin. Radiation Research *5*, 372–381.

KERTESZ, Z. I., SCHULZ, E. R., FOX, G., and GIBSON, M. 1959. Effects of ionizing radiations on plant tissues. IV. Some effects of gamma radiation on starch and starch fractions. Food Research *24*, 609–617.

KRAYBILL, H. F., READ, M. S., HARDING, R. S., and FRIEDEMANN, T. E. 1960. Biochemical alteration of milk proteins by gamma and ultraviolet radiation. Food Research *25*, 372–381.

LICCIARDELLO, J. J., NICKERSON, J. T. R., PROCTOR, B. E., and CAMPBELL, C. L. 1959A. Storage characteristics of some irradiated foods held at various temperatures above freezing. I. Studies with chicken meat and sweet potatoes. Food Technol. *13*, 398–404.

LICCIARDELLO, J. J., NICKERSON, J. T. R., PROCTOR, B. E., and CAMPBELL, C. L. 1959B. Storage characteristics of some irradiated foods held at various temperatures above freezing. II. Studies with pork sausage and scallops. Food Technol. *13*, 405–410.

LIM, E., YEN, J., and FENTON, F. 1959. Effect of irradiation on quality of ground pork and of cooking conventionally and electronically. Food Research *24*, 645–658.

LLOYD, N. E., MILNER, M., and FINNEY, K. F. 1957. Treatment of wheat with ionizing radiations. I. Some effects of x-rays on gluten and gluten sols. Cereal Chem. *34*, 55–62.

Lück, H., Schillinger, A., and Kohn, R. 1960. Effect of electronic radiation on carrot powder. Z. Lebensm. Untersuch. U. Forsch. *111*, 307–318.

Mateles, R. I., and Goldblith, S. A. 1958. Some effects of ionizing radiations on gelatin. Food Technol. *12*, 633–639.

Mehrlich, F. F. 1961. Progress is being made in food irradiation. Food Processing *22*, No. 5, 40–43.

Miyauchi, D. T. 1960. Irradiation preservation of Pacific Northwest fish. I. Cod fillets. Food Technol. *14*, 379–382.

Mullins, W. R., and Burr, H. K. 1961. Treatment of onions with gamma rays. Effects of delay between harvest and irradiation. Food Technol. *15*, 178–179.

Naik-Kurade, A. G., Livingston, G. E., Francis, F. J., and Fagerson, I. S. 1959. Effects of cathode ray and gamma ray irradiation on some organic acid-carbohydrate systems. Food Research *24*, 618–632.

Nicholas, R. C., Meiske, D. P., Jones, M. F., Wiant, D. E., Pflug, I. J., and Jones, E. M. 1958. Radiation-induced changes in bread flavor. Food Technol *12*, 52–54.

Pearson, A. M., Bratzler, L. J., and Costilow, R. N. 1960A. The effects of pre-irradiation heat inactivation of enzymes on palatability of beef and pork. Food Research *25*, 681–686.

Pearson, A. M., Bratzler, L. J., and Gernon, G. D., Jr. 1960B. The effects of pre- and post-enzyme inactivation storage on irradiated beef and pork roasts. Food Research *25*, 687–692.

Salunkhe, D. K. 1957. Histological and histochemical changes in gamma-irradiated lima beans, *Phaseolus lunatus*. Nature *179*, 585–586.

Salunkhe, D. K., Gerber, R. K., and Pollard, L. H. 1959. Physiological and chemical effects of gamma radiation on certain fruits, vegetables, and other products. Proc. Am. Soc. Hort. Sci. *74*, 423–429.

Shea, K. G. 1958. Food preservation by radiation as of 1958. Food Technol. *12*, 6–16.

Stadelman, W. J., and Wise, R. G. 1961. Tenderness of poultry meat. I. Effect of anesthesia, cooking, and irradiation. Food Technol. *15*, 292–294.

Webb, N. L., Rutherford, B. E., and Wiant, D. E. 1961. Effect of ionizing radiation on cakes and biscuits made with milled irradiated wheat. Food Technol. *15*, 386–388.

Wertheim, J. H., Roychoudhury, R. N., Hoff, J. E., Goldblith, S. A., and Proctor, B. E. 1957. Radiation preservation of milk and milk products. J. Agr. Food Chem. *5*, 944–959.

CHAPTER 15

Chemical and Osmotic Processing Methods

INTRODUCTION

Osmotic and chemical methods are used to preserve a wide variety of foods. Some of these processes have as their primary goal the improvement of the appearance, flavor, texture, or physiological effect of the food. Brewing and cheese ripening are examples of such processes. On the other hand, the retention of the characteristics of the fresh material without significant change by the preservative treatment is the aim of many chemical methods, e.g., the addition of benzoic acid to beverage concentrates. Texture changes may be of minor importance in some of these processes, while in others the texture is drastically altered.

The preservative effect itself—mechanisms by which original texture is preserved—will not be discussed here. Instead, the discussion will be directed toward texture changes occurring during the processing as well as to spoilage reactions which result from a failure of the preservative effect. A large number of chemical preservatives used in small quantities and having some specific inhibitory action on the life processes of micro-organisms do not themselves have any detectable effect on the texture of foods. Among these compounds are chlortetracycline, nisin, benzoic acid, propionic acid, and propylene oxide. Because they lack a direct effect on texture they will not be discussed further in this chapter.

Some of the most common of the "chemical" preservative methods rely on the abstraction of the most effective nutrients for micro-organisms by controlled microbial growth. Frequently, conditions unsuitable for further microbial proliferation are established as a result of substances secreted into the medium during the controlled growth phase. For instance, the pH of many fermented vegetables is too low to permit growth of most species of bacteria as a result of the elaboration of lactic acid early in the processing. Some very similar procedures are applied with the sole aim of changing the flavor of the product, yet they frequently cause alterations (desirable or undesirable) in the texture of the food.

Preservation by immersion in solutions of organic acids, usually acetic, or of salt is common. The osmolality of the aqueous phase may also be raised to inhibitory levels by soaking foods in concentrated solutions of sucrose or other sugars. Large pieces of meat are frequently treated by packing them in mixtures of sugar, salt, nitrates, and nitrites, the latter compound having a specific protective action. All of these treatments would be expected to have pronounced effects on the texture of the foodstuff.

PROCESSING METHODS AND THEIR EFFECTS

Cheese Preparation and Ripening

The step common to all cheese preparation methods is the formation of casein clots or precipitates from milk. Usually this precipitation is the result of the action of rennin, an enzyme, and acid. The acid is, in practice, formed *in situ* by more or less pure bacterial cultures, although various authors (Deane and Hammond 1960) have suggested the use of added chemical acidogens such as lactide and glucono-delta-lactone. The bacterial cultures also provide special flavors which have come to be expected and desired in cheese.

Cottage cheese, probably the simplest cheese to make and certainly the most popular one, is essentially the drained curd prepared as described above although often some cream is mixed into it to improve the texture. Acid is formed in skim milk by lactic acid bacteria of the *Streptococcus lactis* type. Usually, rennet is also used. The firmness of the curd is at least partially a function of the pH. Size of the clots, which affects perceived texture, is governed by the rate at which the pH is reduced, i.e. the rate of acid development. The most serious texture defect is lack of firmness in the curd and this is due to insufficient acid production. In extreme cases, the curd may fail to develop at all. The curd is finally cooked to a temperature of about 130° to 140°F. Overcooking causes a mealy texture of the curd.

Emmons *et al.* (1960) examined samples of cottage cheese coagulum made with 17 different cultures of lactic acid bacteria. Curd strength measurements were made with the Cherry-Burrell curd tension meter and knife (Anon. 1955). They found that a close relationship existed between the pH of the curd and the curd strength. Regardless of their activity, the cultures showed essentially the same relation between curd strength and decrease in pH after coagulation. The 17 cultures varied widely in their relations between curd strength and titratable acidity of whey, and between pH of curd and titratable acidity of whey. In other words, pH rather than titratable acidity seemed to be the factor influencing texture.

In one texture defect of cottage cheese, the curd particles may become coated with a viscous slimy film due to proteolytic activity of contaminating organisms such as *Alcaligenes viscosus*.

Cream and Neufachatel cheeses are manufactured similarly to cottage cheese and suffer from many of the same texture problems.

Ripened cheeses constitute a very numerous class containing such diverse examples as Cheddar, Swiss, blue, Limburger, Camembert, and Parmesan. The drying which occurs during the storage period requisite for curing gives these varieties a firmer texture than their simpler counterparts unless the ripening organisms exude proteolytic enzymes. Limburger, hand cheese, and Camembert are varieties which depend for their characteristic soft or slimy texture on the partial digestion and liquefaction of denatured casein by micro-organisms. In general, such cheeses develop a rind or surface layer of bacteria and molds in response to traditional processing methods. Camembert, as an example, is covered at the termination of ripening by a felt-like layer of *Penicillium camemberti* and other molds which reduce the acidity of the cheese and produce protein- and fat-digesting enzymes. These enzymes diffuse inward and create a soft, buttery texture by digesting the protein and making it more soluble in the approximately 48 per cent of water which is present. The slime or "smear" on the surface of limburger contains yeasts which consume acid and make conditions favorable for the growth of *Bacterium linens* and other micro-organisms which produce extracellular proteases. There are no objective methods for evaluating texture of these cheeses.

Buttermilk Preparation

Buttermilk is prepared from pasteurized skim milk by inoculating it with about 0.5 to 1 per cent of a butter starter culture and then ripening it for 12 to 16 hours until it reaches a titratable acidity of 0.70 to 0.90 per cent. The textural qualities are due to the presence of a suspended fine precipitate of casein in whey. The size of the particles of casein, which largely determines the viscosity as well as the stability of the suspension, is due to the rate and extent of acid production as well as to many other poorly understood factors. It is common practice in some parts of the country to add a small amount of butter flakes to the milk mainly for the improvement of appearance although the presence of particles of contrasting texture may be pleasant to the consumer.

Fermentation of Fruits and Vegetables

Green beans, corn, okra, many kinds of peppers, and other vegetables as well as green tomatoes have been preserved by fermentation methods. However, cucumbers, cabbage, and olives are the foods most commonly processed in the United States by these methods. Perhaps the original purpose of vegetable fermentation techniques was to preserve the material, but this function is now less important than securing the flavor changes brought about by the microbiological activity. In other parts of the world having less advanced food processing industries, preservation is still an important goal of the fermentation methods.

The preservative effect of fermentation techniques is due both to the metabolites given off by the micro-organisms and to the decomposition of the simple carbohydrates which might otherwise serve as substrates for spoilage bacteria, yeasts, and molds. It is important that the predominant species acting during the fermentation not attack the higher carbohydrates, the proteins, or the lipids for such activity, if very extensive, adversely affects the texture, flavor, and appearance of the food. It can be stated as a general rule that proteolytic, amylolytic, and cellulolytic activities are never desirable and can lead to the development of atypical textural qualities if present in significant amounts.

Salt concentration, temperature, and oxygen tension are the conditions which determine the species of organisms predominating in a food fermentation. Pure cultures of bacteria are not used in practice, although the effects of their use have been studied experimentally by Pederson and Albury (1961). Salt has a selective inhibitory action which permits the growth of certain lactic acid-producing bacteria, a few yeasts (particularly *Debaryomyces spp.*) and numerous kinds of molds. Many spore-forming and non-spore-forming aerobic bacteria and some spore-forming anaerobic bacteria are strongly inhibited by levels of sodium chloride which are only slightly inhibitory for lactic bacteria. The organisms which grow best in brines do not elaborate significant amounts of extracellular enzymes attacking texture-affecting structures. The acids and the salt present in the fermented product do not, in the concentrations normally encountered, cause appreciable deterioration of the structural components of the food.

Sauerkraut.—In preparing this typical fermented foodstuff, cabbage is harvested, allowed to wilt for a day or two, and then shredded. From 1.5 to 2.5 lbs. of salt are added for each 100 lbs. of cabbage, and the mixture is allowed to ferment, preferably at 60° to 75°F. The fermentation of the cabbage is initiated by *Leuconostoc mesenteroides* and completed by *Lactobacillus plantarum* and *Lactobacillus brevis*. After approximately seven days (the time depending upon the temperature, salt concentration, and original inoculum), the acidity in the aqueous phase has reached 1.5 to 2.0 per cent, measured as lactic, and the fermentation is substantially complete. Three to four weeks may elapse before the best textural and flavor qualities are developed.

The desired texture is not much different from that of the wilted cabbage, being slightly crisp or crunchy, and not soft, mushy, or leathery. Evidently only slight modification of the native structural elements of cabbage occurs as a result of the fermentation reactions and the metabolites released into the liquid. The texture quality of wilted cabbage is due chiefly to cellulose fiber structures in the cell wall. These are relatively resistant to attack by chemicals or enzymes. However, some bacteria and molds do secrete cellulolytic enzymes and texture retention in kraut depends upon preventing growth of these organisms in the fermentation medium.

If air pockets, uneven distribution of salt, or high temperatures occur, soft kraut may be formed as the result of the growth of undesirable types of micro-organisms. In extreme cases, complete breakdown of the cabbage shreds is observed. Slimy or ropy kraut is a texture defect caused by aberrant growth of *Lactobacillus plantarum*. At relatively high temperatures this organism becomes encapsulated and, in masses, forms the slime which coats the kraut.

The importance of salt concentration to the texture of the finished kraut was emphasized by a study of Pederson and Albury (1954). They found that all kraut samples made with one per cent salt were too soft. Increasing tenderometer readings were observed as the salt content was raised in steps to about two and one-half per cent. There was not much difference in texture due to incubation at 45.5°, 64.4°, 73.4°, 89.6°, or 98.6°F. except at the two highest temperatures where the texture tended to be tough. Pederson and Albury did not speculate on the reasons for the increased toughness.

Producers usually judge the texture quality of kraut on the basis of subjective examinations. Holfelder (1957) described the use of a texture-ometer for measuring the consistency of raw sauerkraut. The instrument was found to yield values well correlated with panel estimates of the consistency of kraut samples.

Cucumbers.—According to one authority, cucumber pickles can be classified as follows: (1) Fermented pickles—(a) overnight dills; (b) genuine dills; (c) salt stock from which sweet, sour, and mixed pickles are made. (2) Unfermented fresh pasteurized pickles. (3) Sweet, sour, and mixed pickles, and relish made from fermented pickles. The preparation of fermented pickles resembles in many respects the fermentation of sauerkraut. There are two kinds of fermentations which are in common use, the low salt and the high salt methods. In the former, cucumbers are fermented in a brine of eight per cent (30° salinometer) concentration, or less. The salinometer test value is raised two degrees weekly until the brine reaches 50° salinometer after which it is raised one degree weekly until it reaches 60° salinometer. In the high salt method, the pickles are fermented in a 40° or higher brine, and the concentration is raised two degrees weekly until the brine tests 60° salinometer. At the end of these times it is expected that the fermentation will be completed with a final acidity of about 0.4 to 0.7 per cent as lactic. The predominant organism during the fermentation is *Lactobacillus plantarum*.

According to Nicholas and Pflug (1960),

The ideal pickle is crisp; it parts with a distinct snap when bitten and the chewing is accompanied by a definite crunching. The texture of whole pickles can be measured by the Magness-Taylor fruit pressure tester.

These investigators used a $^5/_{16}$ inch head on the tester and reported results as pounds of pressure. Pickles testing more than 10 pounds were regarded by them as having excellent texture; values of 5 to 10 pounds were suggestive of deterioration. Pickles testing less than five pounds were thought to be unsatisfactorily soft. Variation from jar to jar of the same experimental variable was found to be about 2 to 3 pounds.

The chemical changes undergone by the structural elements of the pickle during fermentation are poorly understood. Most investigators

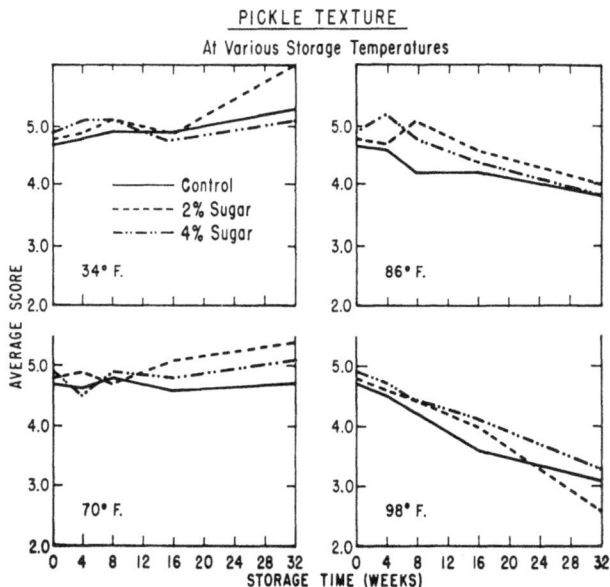

FIG. 27. EFFECT OF STORAGE TEMPERATURE AND SUGAR CONCENTRATION ON PICKLE TEXTURE

have restricted themselves to studies of gross changes in texture. Pangborn et al. (1959) investigated the effects of sugar, storage time, and temperature on processed dill pickle quality. Shear press values were significantly correlated with a panel's texture scores ($r = +0.71$) and with color scores ($r = -0.70$). Texture quality decreased as the storage temperature was increased. Fig. 27 shows the interaction of sugar concentration and storage temperatures. Firmness increased during storage at 34°F., while softening occurred at 86° and 98°F. There was very little texture change during storage of the pickles at 70°F. for up to 32 weeks. They attributed these observations to changes in the pectic substances and to a cellular change brought about by replacement of intercellular air by fluid.

Bell and Etchells (1960) showed that use of increasingly higher salt concentrations gave correspondingly higher values for cucumber firmness in accordance with first-order reaction kinetics.

Nicholas and Pflug (1960) also studied the effect of storage temperature on pickle quality. At 40°F. all pickles remained in excellent condition throughout the test period of 388 days. Similar results were observed at 72°F. Deterioration occurred at 86°, 90°, and 100°F. Loss of crispness was usually observed first in the seed cavity which lost body texture altogether; the flesh failed to crunch upon chewing; in the final stages, the skin tears readily, having lost its characteristic texture. These changes may have been due to a gradual hydrolysis of the cementing substances of the middle lamellae by the acidic medium.

Potassium aluminum sulfate has been used as a texture-improver in pickles. When added to the finished product, it firms the pickle, making it crisper. The trivalent aluminum ion exhibits similar effects in other foods. In the case of pickles, it probably acts by complexing with the pectic substances, particularly those in the middle lamellae.

In summary, it can be said that pickle defects can originate during fermentation or in storage. In the former instance, the texture faults are due to secretion of cellulolytic or pectinolytic enzymes into the fluid by contaminating micro-organisms or other sources. Storage changes are probably usually due to a gradual non-enzymatic hydrolysis of the cementing substances between the cells.

Olives.—The texture of olives depends somewhat on the variety. Different varieties typically contain different quantities of oil, which has a pronounced effect on the texture. Before processing, the olives are stored in brines of 3 to 10 per cent concentration. Salt levels above ten per cent cause shriveling of the olives because of withdrawal of fluid from the cells by osmotic action.

Most olives except Sicilian-style olives are treated with solutions of about 0.25 to 2 per cent sodium hydroxide in order to remove the bitter principle and to develop a darker color. Unfortunately, there do not seem to be any publications describing the effects of this treatment on the texture of the fruit. However, this concentration of alkali is sufficient to cause changes in the characteristics of the structural components of the olive. It would be valuable to have quantitative data on the changes in the physical properties of the olive pulp as related to time, temperature, and concentration of alkali.

After the lye has been washed from the fruit, it is placed in salt solution and allowed to ferment for several days during which time the concentration of brine is steadily increased by additions of salt. Some fruits may increase in size or "plump." This change yields a somewhat firmer texture. Spoilage reactions due to enzymes secreted by contaminating species of bacteria can cause the olives to become soft, mushy, or of uneven texture.

Other Fruits and Vegetables.—The desirable texture of onions and cauliflower is said to be destroyed by fermentation. Consequently, they are best preserved in concentrated brines, say 80° salinometer. Acetic acid may be added to the brined material without causing undesirable softening. Green tomatoes do not yield enough soluble carbohydrate to allow an inhibitory amount of acid to be developed by fermentation. If it is desired to ferment this fruit, sugars may be added to support a more extensive fermentation. Green beans, okra, corn, sweet red peppers, and Zucca melon have been successfully treated by lactic fermentations with retention of desirable texture during processing and storage.

Candying

Pineapples, cherries, pears, peaches, citron, orange peel, lemon peel, and ginger are commercially prepared by infusing them with saturated or nearly saturated sugar syrups. Often the raw material for candying is fruit which has been preserved in sulfite brines for long periods. The usual procedure is to immerse the product in a series of hot sugar syrups of increasing concentration. One result of this process is a thorough cooking of the material. Cell semipermeability is lost, of course, and turgor pressure is not a factor in the texture of the candied food. Sterling and Chichester (1960) discussed the distribution of sugar in plant tissues cooked in syrups of considerably lower concentration than those reached in candying and glacéing processes. It appears that dextrose is concentrated in the cell wall by adsorption. Although Sterling and Chichester do not speculate on the textural effects of this adsorption, it probably firms the cell walls with consequent development of a harder, tougher texture. Presence of a viscous sugar syrup in the cell lumens and in the remnants of the shrunken intercellular spaces would also seem to contribute to the firmness of the product. The original cellulosic, ligneous, and pectinous structures probably remain substantially intact and affect the texture in the usual manner. Use of sucrose syrups under certain conditions can lead to crystallization of sugar on the surface of the piece, yielding the appearance and texture typical of crystallized ginger. The internal texture is also likely to have a granular component under these conditions.

Curing of Meats

Curing refers to the treatment of meats with combinations of sodium chloride, sodium nitrate, sodium nitrite, sugar, and other flavors and preservatives. Preservation of the meat and modification of its flavor and texture are the purposes of this treatment. Meat may be rubbed with, or packed in, the dry ingredients, in which case it is said to be dry-cured, or it may be immersed in, or injected with, a solution of the curing substances. Comminuted meats may be mixed with the chemicals.

A dry cure creates a firmer texture not only by encouraging dehydration and increasing the osmotic pressure of the tissues, but also by coagulation of the proteins. The first two changes are reversed by cooking procedures, but texture changes due to premature precipitation of the proteins may persist through ordinary preparation techniques. The moisture content of the meat is increased by injection of pickle and the meat can be made softer if the increased moisture content is carried through to the finished product.

Smoking.—Smoking is used alone and in combination with other methods to extend the storage life of certain meats and to improve their flavor and texture. Heat, volatile chemicals generated by combustion, and drying contribute to the observed effects. The cooking and drying reactions are essentially the same as those covered in the appropriate sections of this book. Among the chemicals present in wood smoke and condensing on the surface of foods being treated by this process are pyroligneous acid (mostly acetic acid), formaldehyde, creosote, and phenols. Probably the direct influence of these compounds on texture is rather minor, although some coagulation of proteins near the surface of the meat may result. This effect is doubtless obscured by the cooking process.

SUMMARY

Fermentation processes for preserving fruits and vegetables generally leave the principal structural elements intact although the semipermeability of the cell membrane is destroyed with loss of the contribution of cell turgor to texture. Loss of desirable textural qualities during fermentation is usually the result of secretion into the medium of pectinolytic and cellulolytic enzymes by contaminating bacteria. Softening during storage may occur, especially at temperatures above about 80°F., as the cementing substance of the middle lamellae are attacked by the acids developed during fermentation or added as flavoring ingredients.

Candying and glacéing cause an increase in rigidity, probably by decreasing the mobility of polymeric molecules by dehydration. Curing of meats by chemical and osmotic means can result in a toughening due to dehydration of the tissues and denaturation of the proteins, or a softening

due to swelling and partial depolymerization, depending upon the method
of curing which is used. Smoking is essentially a cooking and dehydrating
process, so far as texture changes are concerned.

BIBLIOGRAPHY

ANON. 1955. Curd tension meter. Cherry Burrell Corp. Bull. *1P 5132-M.*

BELL, T. A., and ETCHELLS, J. L. 1960. Influence of salt on pectinolytic softening of cucumbers. Food Research *25*, 84–90.

BELL, T. A., ETCHELLS, J. L., and JONES, I. D. 1955. A method for testing cucumber salt-stock brine for softening activity. U. S. Dept. Agr. ARS 72-5.

BRADY, D. E., SMITH, F. H., TUCKER, L. N., and BLUMER, T. N. 1948. Characteristics of country-style hams as related to sugar content of curing mixture. Food Research *14*, 303–311.

BREKKE, J. E., and SANDOMIRE, M. M. 1961. A simple, objective method of determining firmness of brined cherries. Food Technol. *15*, 335–338.

BULLIS, D. E., and WIEGAND, E. H. 1931. Bleaching and dyeing Royal Anne cherries for maraschino or fruit salad use. Oregon State Coll. Agr. Expt. Sta. Bull. 275.

DEANE, D. D., and HAMMOND, E. G. 1960. Coagulation of milk for cheese making by ester hydrolysis. J. Dairy Sci. *43*, 1421–1429.

DEMAIN, A. L., and PHAFF, H. J. 1957. Cucumber curing: Softening of cucumbers during curing. J. Agr. Food Chem. *5*, 60–63.

EMMONS, D. B., PRICE, W. V., and TORRIE, J. H. 1960. Effects of lactic cultures on acidity and firmness of cottage cheese coagulum. J. Dairy Sci. *43*, 480–490.

ETCHELLS, J. L., BELL, T. A., and WILLIAMS, C. F. 1958. Inhibition of pectinolytic and cellulolytic enzymes in cucumber fermentation by Scuppernong grape leaves. Food Technol. *12*, 204–208.

HAMILTON, I. R., and JOHNSTON, R. A. 1961A. Studies of cucumber softening under commercial salt-stock conditions in Ontario. 1. Incidence and pattern of activity of pectolytic enzymes. Appl. Microbiol. *9*, 121–127.

HAMILTON, I. R., and JOHNSTON, R. A. 1961B. Studies of cucumber softening under commercial salt stock-conditions in Ontario. 11. Pectolytic micro-organisms isolated. Appl. Microbiol. *9*, 128–135.

HOLFELDER, E. 1957. Experiences in the use of the texturemeter for measurements of the consistency of raw sauerkraut. Ind. Obst. u. Gemüseverwert. *42*, 39–41.

HUNT, W. E., SUPPLEE, W. C., MEADE, D., and CARMICHAEL, B. E. 1939. Qualities of hams and rapidity of aging as affected by curing and aging conditions and processes. Maryland Univ. Agr. Expt. Sta. Bull. *428.*

KEMP, J. D., MOODY, W. G., and GOODLETT, J. L. 1961. The effects of smoking and smoking temperatures on the shrinkage, rancidity development, keeping quality, and palatability of dry-cured hams. Food Technol. *15*, 267–270.

KERTESZ, Z. I. 1951. The Pectic Substances. Interscience Publishers, New York.

LABIE, C. 1958. The use of phosphates in the preservation of meat products. Rec. Med. Vet. Ecole d'Alfort *134*, 133–138.

MCCREADY, R. M., and McCOMB, E. A. 1954. Texture changes in brined cherries. Western Canner Packer *46*, No. 12, 17–19, 24.

MILLER, R. C., and ZIEGLER, P. T. 1936. How aging affects the distribution of salt in cured hams. Food Inds. *8*, 121–130.

NICHOLAS, R. C., and PFLUG, I. J. 1960. Effects of high temperature storage on the quality of fresh cucumber pickles. Glass Packer *39*, No. 4, 35, 38–39, 65.

NIVEN, C. F., JR. 1960. Factors influencing quality of cured meats. *In* The Science of Meat and Meat Products. W. H. Freeman and Co., San Francisco, Calif.

PANGBORN, R. M., VAUGHN, R. H., YORK, G. K., III, and ESTELLE, M. 1959. Effect of sugar, storage time, and temperature on dill pickle quality. Food Technol. *13*, 489–492.

PEDERSON, C. S., and ALBURY, M. N. 1954. The influence of salt and temperature on the microflora of sauerkraut fermentation. Food Technol. 8, 1–5.

PEDERSON, C. S., and ALBURY, M. N. 1961. The effect of pure-culture inoculation on fermentation of cucumbers. Food Technol. 15, 351–354.

PEDERSON, C. S., and KELLY, C. D. 1952. Quality of commercial sauerkraut. N. Y. State Agr. Expt. Sta. Bull. 613.

POLLARD, A. 1958. Fermentation of diluted concentrates. In Int. Fed. Fruit Juice Producers' Symp., Bristol, 1958, 351–360.

ROSS, E., YANG, H., and BREKKE, J. E. 1958. Preliminary report on the brined cherry project, 1958 season. Wash. State Coll. Agr. Expt. Stas. Circ. 340.

STEELE, W. F., and YANG, H. Y. 1960. The softening of brined cherries by polygalacturonase, and the inhibition of polygalacturonase in model systems by alkyl aryl sulfonates. Food Technol. 14, 121–126.

STERLING, C., and CHICHESTER, C. O. 1960. Sugar distribution in plant tissues cooked in syrup. Food Research 25, 157–160.

YANG, H. Y., STEELE, W. F., and GRAHAM, D. J. 1960. Inhibition of polygalacturonase in brined cherries. Food Technol. 14, 644–647.

Spontaneous Changes Affecting the Texture of Foods

CHAPTER 16

Effects of Physical Changes

INTRODUCTION

Many of the changes which occur during storage and result in the textural deterioration of foods are not the consequence of any chemical reaction, either enzymatic or non-enzymatic. Such phenomena as "blooming" in chocolate, graining in caramels, softening of cookies and crackers, freezerburn, weeping of gelatin desserts, development of stickiness or sandiness in hard candies, and staling of baked products are typical of changes in this category. Most of these changes are due to one of the following causes:

(1) Changes in moisture content due to evaporation or absorption or to translocation of moisture within the food.

(2) Redistribution of solutes or fatty substances within the food.

(3) Crystallization in the broadest sense, including the formation of micelles in aggregates of starch and cellulose molecules.

(4) Separation of emulsions.

(5) Melting, with or without subsequent soldification.

(6) Syneresis of gels.

All of the changes result from a metastable condition of the food, frequently existing from the moment of preparation, but occasionally the result of the unexpected occurrence of unfavorable environmental conditions.

Although gross changes in texture resulting from mechanical damage (e.g., the breaking of crisp materials into small pieces) could be considered as falling under the present heading, they are not of sufficient theoretical interest to warrant discussion in this book.

CRYSTALLIZATION

Many structural components of foods contribute desirable physical qualities only when they are in an amorphous or a hydrated state. For example, hard candies are glassy compositions, i.e., amorphous, and the starch of bakery goods is a highly hydrated substance, in the preferred forms of the foods. Crystallization of the sucrose in hard candy and retrogradation of starch (which can be considered a kind of crystallization involving only parts of the molecules) in baked products destroy the characteristic textures of these foods. Such processes, though strictly physical in nature, are economically important storage factors.

Sugar bloom in chocolate is due to the formation on the surface of the piece of palpable crystals of sucrose.   These crystals give a gritty or sandy texture to the candy.   The sucrose used as a sweetener in all "eating" varieties of chocolate is ordinarily present as very fine particles, certainly less in any dimension than the 50 microns or so required for particles to become perceptible to the touch.   Since the moisture content of chocolate is of the order of 1 or 2 per cent and most of it is adsorbed on insoluble materials, diffusion of the sugar is not feasible and growth of crystals never occurs.   However, under certain conditions of storage, free water can be present in sufficient quantity to permit the crystals of sucrose to develop to perceptible size through a process of solution and re-precipitation.   In addition to the obvious, but rather unlikely, occurrence of accidental wetting of the candy, the surface can be moistened by the condensation of water on cold pieces which have been removed to a warmer and more humid atmosphere.   For this reason, storage of chocolates at refrigerator temperatures must be carried out with a certain degree of caution, and frequent transfers from refrigerated containers to room temperature areas is definitely contra-indicated.   Once sugar bloom has formed, no technique other than complete re-processing of the piece can remove it.

Neville *et al.* (1950) indicate that sugar bloom can be prevented by the use of a minimum of 34 per cent fat in the chocolate.   Evidently, this amount of fat is necessary to completely isolate each sucrose particle from any potential aqueous phase.   These authors also include proper control of moisture content and careful handling and storage as necessary steps for preventing storage bloom.

Crystallization in table syrups is very undesirable.   In corn syrups, this change is prevented by keeping the dextrose content below a certain level.   In cane and maple blends, the sucrose concentration is kept low enough that normal fluctuations in temperature will not cause precipitation.   The problem is most important in honey.   This product contains in excess of 80 per cent solids, mostly glucose and fructose.   The glucose has a tendency to precipitate as the crystalline monohydrate, causing the honey to have a grainy rather than a smooth consistency.   Precipitation is inhibited to some extent by protective colloids native to the product and by the other carbohydrates which are present.

Crystal size is all important in determining the undesirability of graining in honey.   In fact, grained honey is deliberately produced as an article of commerce.   Here the crystal size is kept very small and uniform so that an undesirable gritty sensation is not imparted to the consumer.   The trick of the operation is to secure many very minute foci for crystallization.   The presence of a relatively few large crystals or of a wide range of crystal sizes will lead to growth of the larger crystals at the expense of the smaller, with ultimate development of graininess. This texture defect can be reversed simply enough by heating the honey to a temperature high enough to dissolve all of the crystals.

One of the texture defects which can develop in ice cream is "sandiness," which is caused by the growing of lactose crystals or ice crystals to a size which is large enough to be tactually detectable. Lactose is a relatively poorly soluble sugar and the temperature at which ice cream must be stored is inconsistent with the concentration of lactose initially present in the aqueous phase of the mix. Consequently, lactose crystals are formed in all samples of ice cream after a short storage period. Ideally, however, they are of imperceptible size when precipitated and remain so throughout the storage period. If the time of storage is exceptionally long, or if the temperature of storage fluctuates, development of sandiness may be observed. This is due, of course, to redistribution of the lactose from the smaller to the larger crystals. In order for the lactose to diffuse to the crystallization foci, water in the liquid state must be present. This requirement would suggest that a way to prevent growth of sugar crystals in ice cream would be to store the product at a temperature sufficiently low to solidify all of the aqueous phase. Perhaps a more practical method was suggested by Nickerson (1954) who advocated the addition of crystal "seeds" so that the excess lactose would be precipitated at an early stage of the chilling process in the form of crystals less than (e.g.) ten microns in size. In ice cream treated in this manner, development of sandiness is much retarded but even here it will occur eventually.

Ice crystals can form in ice cream leading to a texture defect. As in the lactose situation, growth occurs during storage, but, unlike lactose, water can redistribute in the vapor phase and the ice crystals can grow by accretion even though no liquid water is ever present during storage. In this case, as in the others described in the present section, the size distribution of the crystals which are initially deposited is of paramount importance in determining the rate of growth and the time of appearance of palpable crystals.

The development of graininess in hard candy and caramel is another example of texture deterioration due to sucrose crystallization. Hard candies are essentially highly supersaturated syrups of sucrose and glucose, usually with some fructose. In order to promote smooth texture, it is desirable that these products be entirely free of crystals. Since the sugar concentrations are much above the saturation level, a metastable system with a strong tendency to precipitate crystals is formed. As long as the moisture level remains on the order of 1 to 4 per cent, crystallization is prevented because the sucrose molecules cannot diffuse into the ordered positions required for forming a crystal lattice. However, a moisture content much above this level permits a rapid crystallization which ultimately destroys the characteristic texture of the candy. The usual sequence of events in storage deterioration is the adsorption of moisture on the surface of the piece with partial solution of sugars, then graining becomes apparent, and water is released to promote the same changes in the underlying layer. The whole piece of candy can become a friable mass of minute crystals through the operation of this mechanism. The only satisfactory preventative measure is to protect the candy from contact with humid atmospheres.

Fondant, a plastic two-phase system of sugar crystals and saturated syrups which forms the basis for many types of candies and icings, undergoes texture changes attributable to crystal alterations. Fondant, when first formed, is apt to be very stiff. After it stands for some time, it softens. Karácsony and Pentz (1955A and 1955B) and Maczelka (1956) found that softening resulted from an increase in the average crystal size with a resultant decrease in the total surface area in the mass. The reduction in surface area reduces the forces binding the liquid so that it flows more freely. It is also possible that the liquid actually contains more dissolved substances when the average crystal size is smaller. At any rate, the fluid phase becomes, in effect, less viscous as the particle size of the solid phase becomes larger. If the fondant is stored long enough, the crystal size may go well above 100 microns and the mass may become fluid enough to flow like a syrup. Similar though less extensive changes occur in caramel.

A fairly extensive discussion of practical and theoretical aspects of fondant crystallization can be found in Lowe (1955). She mentions the value of additives such as egg white in controlling crystal growth.

In frozen foods containing large proportions of sugars, such as in some fruit juice preparations (Tressler and Joslyn 1961), the separation of crystalline masses of sugar hydrates has occasionally been observed after long storage. Sucrose hydrates appear as white spherulitic masses which may eventually include the whole mass of the food. They usually require seeding and melt rapidly after the food is thawed. Dextrose hydrate usually appears as small white spheroids scattered throughout the food. It forms readily without seeding and will persist for some time after the food has been brought to room temperature (Cotton *et al.* 1955). The effect these changes would have on texture is readily apparent.

RETROGRADATION OF STARCH AND SIMILAR PROCESSES

Retrogradation of starch is a kind of limited crystallization phenomenon causing texture changes in foodstuffs which must be stored for some time after their starch fraction has been cooked or gelatinized. It produces perceptible effects in dehydrated potatoes and in baked products. It must also have some influence on the texture of canned cooked alimentary pastes and canned rice dishes although such products have not been studied sufficiently to allow a worthwhile discussion to be given here.

X-ray diffraction techniques show that the molecules in ungelatinized starch granules possess a definite crystal-like structure or ordered array. An ordered state is also indicated by the birefringence observed when granules are examined by polarized light. When granules are heated sufficiently in water, birefringence is lost and other changes also indicate that the starch then exists as randomly oriented molecules. The process of gelatinization yields highly hydrated molecules, i.e., "dissolved starch."

When a solution of gelatinized starch is brought back to room temperature, parts of molecules lying adjacent to one another tend to associate through hydrogen bonding and other relatively weak forces, excluding from themselves some of the solvent molecules which have saturated the available bonding sites. This is essentially a crystallization process. It is hindered by the long and devious configuration of the molecules so that a perfect crystal lattice can never be formed. However, the crystallization proceeds far enough to cause insolubilization (precipitation) of molecular aggregates. If branches are present on the molecule, as with amylopectin, retrogradation is still further inhibited, but even then it can occur to a limited extent. Retrogradation tendencies are promoted by high concentrations of starch. In bread doughs, the water content is probably insufficient to allow complete swelling of the starch so that the tendency to retrograde is greater than in more dilute systems.

The role of starch retrogradation in the texture staling of bread crumb has been extensively investigated. The literature up to about 1945 was thoroughly reviewed by Geddes and Bice (1946). As bread stales, the crumb (the interior of the loaf) becomes increasingly tough and hard. Eventually, the crumb becomes rather friable or "crumbly" giving a dry sensation like sawdust in the mouth. These texture changes are accompanied by a decrease in the water absorption capacity of the crumb, changes in the X-ray diffraction pattern, and increases in the crumb opacity, among other alterations. It is reasonably certain that starch retrogradation is a major cause of these crumb-staling phenomena, but it appears that other important factors may also be operative. Starch retrogradation plays no part in crust staling.

Much effort has been devoted to finding methods for reducing the rate of bread staling. The linear component of starch, amylose, retrogrades easily to form insoluble aggregates which cannot be brought back to the solvated condition even by autoclaving. Amylopectin, the branched component, is stable in solution for long periods at room temperature but can be insolubilized by repeated freezing and thawing. Retrograded amylopectin can readily be put into solution by heating. Attempts have been made to produce bread having a reduced firming rate by mixing additional amylopectin into the dough, or by synthesizing flour from vital gluten and amylopectin (containing some amylose, usually). These experiments have been uniformly unsuccessful.

Freezing seems to stop the staling reaction, although temperatures of about 40° to 50°F. accelerate it. Therefore, bread which is to be preserved by frozen storage must be brought through this temperature range as rapidly as possible both during the freezing step and thawing. This approach to bread preservation has been fully discussed in Chapter 12.

Bacterial amylases having high heat stability have been added to the dough, and apparently enough amylase activity survives the baking process to somewhat retard the rate of staling by breaking down the starch molecules. It is difficult to see how the extent of "crystallization" could be reduced by making the molecules smaller; so the bacterial amylases probably exert their softening effect by causing the molecular aggregates to form less rigid and extensive networks. The chief problem encountered with the use of such enzymes is the necessity for providing an ultimate limit to their action so that softening does not continue to an undesirable extent.

The greatest success in increasing the texture-life of bread has been obtained by adding compounds which reduce the speed of retrogradation by forming complexes with the starch molecules. Among the most effective compounds which have been tried are derivatives of polyoxyethylenes and monoglycerides of fatty acids. Generally speaking, the former group of chemicals are the most efficient on a weight basis but they are not permissible additives for bread at this time.

Hopper (1949) used X-ray techniques in combination with compressibility studies to show that the retardation by monoglycerides of firmness changes in bread corresponds qualitatively to a slowing of the rate of starch crystallization. Bradley (1949) demonstrated that polyoxyethylene monostearate had a more pronounced effect on softness at one-half of one per cent than did mixtures of mono- and di-glycerides at the one per cent level. Lecithin had little or no effect on softness. More recent work has shown rather conclusively that di-glycerides exert no improving effect on crumb softness.

Although the tenderness of the crumb at the time of ingestion is doubtless of greater importance to the consumer than the over-all softness of the loaf at the time of purchase, the buyer is guided more by the latter factor—of necessity.   Practically all data on staling rates refers to slices or other crumb pieces.   Very recently a "squeeze-test" designed to measure the factors governing consumer evaluation of the texture of a wrapped loaf has been designed by Dalby and Hill (1961).   See Fig. 28.

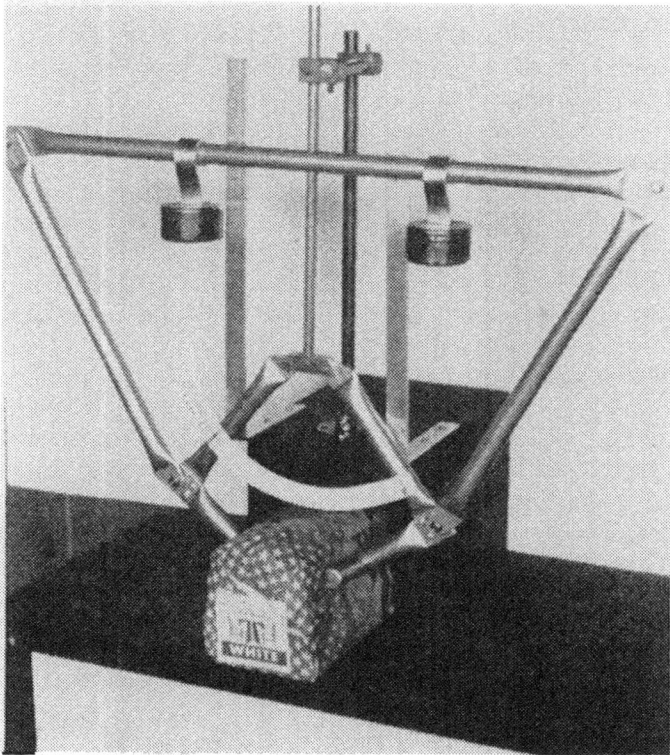

Courtesy G. Dalby

FIG. 28.   THE DALBY-HILL SQUEEZE-TESTER FOR BREAD

Cellulose as well as starch can undergo a kind of crystallization (Battista 1950) in which hydrogen-bonds to water molecules are replaced by intramolecular bonds.   These changes are apparently unimportant and may not occur in foods of medium to high moisture contents.   Cellulose crystallinity can increase during storage of dehydrated samples (Shimazu and Sterling 1961) with a consequent increase in the elasticity of the rehydrated cell wall.   The substance of higher crystallinity swells poorly.   The more highly crystalline (and elastic) the cell wall, the poorer the volume of reconstitution, according to Sterling and Shimazu (1961).

It may be that increases in crystal forms, or bonding together of protein molecules by salt- or hydrogen-bonds is the explanation for some of the decrease in hydration capacity of freeze-dried meats during storage (Hamm and Deatherage 1960).

DIFFERENTIAL MELTING

There are only a few cases of texture deterioration which can be traced directly to the effects of differential melting of food components during storage. Among the most interesting of these is the development of fat bloom in chocolate.

Fat bloom is characterized by the appearance on the surface of the piece of light-colored patches having a greasy atypical texture. Wide fluctuations in temperature during storage or very long storage times at moderate temperatures can cause fat bloom in any sample of chocolate, but some samples are much more susceptible than others to this texture defect.

The light-colored patches of fat bloom are due to the accumulation at the surface of the chocolate of fat crystals having a different melting point than the cocoa butter as a whole. Cocoa butter consists of a complex of the mixed glycerides of oleic, palmitic, and stearic acids. It is said that these fats (or some of them) are able to crystallize in at least four different forms, each of which is stable within a rather narrow temperature range. Since only one of these forms will be stable at any constant temperature, all of the fat component existing in the other crystal forms will tend to change to the most stable form.

The development of fat bloom has been explained by Neville *et al.* (1950) as follows:

> Crystals of higher melting point will have the best chance to survive and will be to some extent pushed to the outside of the chocolate piece or coating by the liquid fat which expands as it melts. The higher-melting component is soluble in the lower-melting fraction, especially at the higher temperatures, and this dissolved material will deposit on the remaining crystals of the higher-melting component at the surface as the mass cools. The melted portion of the fat will also crystallize and contract upon cooling and the last remaining liquid will perhaps be drawn back into the porous solid mass leaving the bloom above the surface of the chocolate.

Regardless of the correctness of the above explanation, it is very clear that the crystal composition of the solid chocolate, as established by the conditions of the tempering process through which it has passed, determines the relative susceptibility of the material to the development of fat bloom. The storage conditions most conducive to the occurrence of this texture defect seem to be temperatures cycling between a fairly high temperature and a moderate temperature. Duck (1961) described a test for bloom which involved cycling the product for 12 hours at 68°F. and 12 hours at 86°F. Under these conditions, bloom will appear on ordinary samples within 7 to 10 days. Duck found that blooming could be suppressed by adding 0.5 to 5 per cent of a stabilizing ingredient consisting of a mixture of triglyceride esters of fatty acids, principally lauric, myristic, and palmitic in the ratio of 2.0, 1.2, and 2.0.

MOISTURE CONTENT CHANGES

Although absorption of water in the liquid state sometimes occurs as a result of accidental wetting of a food, most of the cases of texture deterioration which can be attributed to unwanted changes in the moisture content of a product are due to the uptake of water vapor. Unless a product is hermetically sealed, some interchange of water vapor between it and the ambient air will always occur. Most foods never reach a completely stable moisture content, even during manufacture. Crisp cereal foods, such as cornflakes, are removed from the heat source before drying is complete, and bread and cake are taken from the oven at a time when water vapor is actively distilling from the product. Hard candy melts are actively changing in moisture vapor relationships when they are discharged from the kettle. A considerable number of other examples of the same condition of moisture instability readily come to mind. The relative humidity and temperature of the atmosphere are constantly changing. As a result of all of these fluctuations, the flow of water vapor to or from the product can be expected to occur at all times except on those rare occasions when the equilibrium relative humidity of the food is exactly the same as the relative humidity of the ambient atmosphere.

The texture of many foods at the time of consumption is relatively insensitive to the movements of water vapor which have occurred during storage. If the food is to be cooked immediately before serving, the preceding moisture changes are largely cancelled out, so far as their effect on texture is concerned.

On the other hand, texture in many hard and crisp foods bears a close relationship to the moisture content. For foods depending for their crispness on a starch and protein network (cookies, ready-to-eat breakfast cereals, snack foods such as potato chips, etc.), a moisture content of less than five per cent appears to be critical in many instances. It is interesting, and probably significant, that moisture contents of about this level are required to establish monomolecular layers in such products (Salwin 1959; Salwin and Slawson 1959).

Shrinkage develops and shrinkage stresses are set up as moisture is removed from gel-like and pastry materials (Görling and Beuschel 1959). Under some conditions, the stresses can cause the product to break up completely. Alimentary pastes, especially in the larger piece sizes, are particularly susceptible to this process. Most macaroni products must be slowly dried by air of gradually decreasing relative humidities so that

TABLE 25

DISTRIBUTION OF SALT IN HAMS DURING CURING AND STORAGE[1]

Number of Days in Cure	Number of Days Held (after Removal from Cure)	Coefficient of Variability of Salt[3]	Salt in Center Subsample, Per Cent	Salt in Outside Subsample, Per Cent
		Dry-cured Hams[2]		
26	0	111.9	1.5	16.3
26	30	12.7	10.3	14.0
28	0	120.8	2.5	13.3
28	30	15.9	8.8	11.9
35	0	70.7	3.1	12.3
35	30	12.2	9.0	10.8
40	0	59.8	3.7	13.5
40	30	8.4	9.7	11.8
42	0	48.7	6.0	15.7
42	30	11.0	10.7	9.7
		Brine-cured Hams[4]		
28	0	83.8	2.1	13.6
28	30	3.2	9.3	9.9
34	0	80.5	3.1	16.9
34	30	33.2	7.0	11.8
35	0	90.9	2.9	18.9
35	30	17.8	10.6	7.5
42	0	73.6	3.3	16.3
42	30	6.4	12.2	10.6

[1] Adapted from Miller and Ziegler (1936).
[2] Curing mixture = 8 lbs. salt, 3 lbs. brown sugar, 3 oz. saltpeter.
[3] Coefficient of variability = $\dfrac{\text{Standard deviation}}{\text{Mean}} \times 100$.
[4] Curing mixture = 8 lbs. salt, 3 lbs. brown sugar, 3 ozs. saltpeter in four gallons of water used for 100 lbs meat.

shrinkage stresses can be relieved by distortion of the product. "Checking" or cracking is a texture defect sometimes observed in crackers and cookies. It is due to the shrinkage stresses set up during the final stages of baking. According to Dunn and Bailey (1928) the conditions that will prevent checking are (in the order of their importance): thorough baking, baking in a humid atmosphere, keeping the biscuit warm as long as possible, thorough mixing, and inclusion of some invert sugar in the formula.

Loss of water can also lead to undesirable texture changes. Obviously, this process is of most importance in products intended to be soft and elastic when consumed. Further declines in moisture content of foods which are crisp and nearly dry rarely occur and, in addition, are of no consequence so far as texture is concerned (although they may be very effective in increasing the rate of rancidity development). A significant contribution to staling of rich cakes is made by drying. Marshmallow confections lose moisture readily with resultant increases in firmness and toughness. Freezerburn in products stored at low temperatures is due to sublimation of water. In most cases, loss of water can be prevented simply by packaging the food in tight-fitting and moisture-impervious containers. This allows the food to establish an equilibrium with the relative humidity of its ambient air with minimum loss of water.

OTHER PHYSICAL CHANGES AFFECTING TEXTURE

Although there are not many examples of texture deterioration due to the redistribution of solutes in foods during storage, a few cases of this kind have been discussed in the literature. Cured meats are products in which a massive diffusion of salts and sugars occurs during processing and storage. The data in Table 25 demonstrate the extent of typical changes in sodium chloride at two locations in hams undergoing dry-cure and brine-cure. It would be expected that the changes in salt concentration would affect the physical properties of the protein fibers which are primarily responsible for the texture of muscle tissue. As stated by Miller and Ziegler (1936), "the curing agent contributes to the final quality of the meat as manifested by its color, flavor, tenderness, and saltiness." It is unfortunate that no attempt has been made to derive a quantitative relationship between the salt concentration and the tenderness (panel or shear).

BIBLIOGRAPHY

BARRETT, C. D. 1961. Temperature and relative humidity determine candy shelf life. Food Processing 22, No. 4, 46–47.

BATTISTA, O. A. 1950. Hydrolysis and crystallization of cellulose. Ind. Eng. Chem. 42, 502–507.

BECHTEL, W. G. 1959. Staling studies of bread made with flour fractions. V. Effect of a heat-stable amylase and a cross-linked starch. Cereal Chem. 36, 368–377.

BLINC, M. 1956. Some observations concerning the effect of beta-amylase on the staling of bread. Brot u. Gebäck 10, 249–252.

BOURNE, E. J., TIFFIN, A. I., and WEIGEL, H. 1960. Interaction of starch with sucrose stearates and other antistaling agents. J. Sci. Food Agr. 11, 101–109.

BRADLEY, W. B. 1949. Bread softness and bread quality. Bakers Digest 23, No. 1, 5–7.

BRADLEY, W. B., and THOMPSON, J. B. 1950. The effect of crust on changes in crumbliness and compressibility of bread crumb during staling. Cereal Chem. 27, 331–335.

BRODY, A. L., and BEDROSIAN, K. 1961. Effect of room temperature vs. refrigerated storage on quality of canned fruit and vegetable products. Food Technol. *15*, 367–370.

BRUNAUER, S., EMMETT, P. H., and TELLER, E. 1938. Adsorption of gases in multimolecular layers. J. Am. Chem. Soc. *60*, 309–319.

CATHCART, W. 1940. Review of progress in research on bread staling. Cereal Chem. *17*, 100–121.

COTTON, R. H., REBERS, P. A., MAUDRU, J. E., and RORABAUGH, G. 1955. The role of sugar in the food industry. *In* Use of Sugars and Other Carbohydrates in the Food Industry. American Chemical Society, Washington.

CRAMER, A. B. 1950. Some problems in the manufacture of hard candy. Food Technol. *4*, 400–403.

DALBY, G., and HILL, G. 1961. A squeeze tester for bread loaves. Paper presented at the 1961 convention of the American Association of Cereal Chemists. Dallas, Texas.

DUCK, W. 1959. Consistency of caramel. Mfg. Confectioner *39*, No. 6, 29–31.

DUCK, W. N. 1961. Bloom inhibited chocolate and method of producing same. U. S. Patent 2,979,407. April 11.

DUNN, J. A., and BAILEY, C. H. 1928. Factors affecting checking in biscuits. Cereal Chem. *5*, 395–430.

EASTON, N. R., KELLY, D. J., BARTRON, L. R., CROSS, S. T., and GRIFFIN, W. C. 1952. The use of modifiers in chocolate to retard fat bloom. Food Technol. *6*, 21–25.

GEDDES, W. F., and BICE, C. W. 1946. The role of starch in bread staling. Quartermaster Corps Report QMC *17-10.*

GÖRLING, P., and BEUSCHEL, H. 1959. Shrinkage stresses in drying of gel-like and pastry materials. Chem.-Ingr. Tech. *31*, 393–398.

GROVER, D. W. 1947. The keeping properties of confectionery as influenced by its water vapor pressure. J. Soc. Chem. Ind. London *66*, 201–205.

HAMM, R., and DEATHERAGE, F. E. 1960. Changes in hydration and charges of muscle proteins during freeze-dehydration of meat. Food Research *25*, 573–586.

HINTON, C. L. 1958. Some aspects of the shelf life of candies. Mfg. Confectioner *28*, No. 7, 13–18.

HOPPER, R. P. 1949. Information concerning the making of quality bread from the standpoint of softness, tenderness, and keeping quality—an evaluation of monoglyceride shortenings in bread. Proc. Am. Soc. Bakery Engineers *1949*, 63–70.

KARÁCSONY, D., and PENTZ, L. 1955A. Further theoretical and practical investigations on the candying and storage of fondant. Élelmézesi Ipar *9*, 236–244.

KARÁCSONY, D., and PENTZ, L. 1955B. Theoretical and practical problems on the candying and storage of fondant. Élemézesi Ipar *9*, 45–52.

KNIGHTLY, W. H. 1958. The use of glyceryl monostearate and related emulsifiers in candy. Mfg. Confectioner *28*, No. 6, 33–34, 36–38, 40, 42.

LOWE, B. 1955. Experimental Cookery from the Chemical and Physical Standpoint. Fourth Edition. John Wiley and Sons, New York.

MACZELKA, L. 1956. Colloid aspects of food chemistry and technology. Élelmézesi Ipar *9*, 199–202.

MAKOWER, B., and DYE, W. B. 1956. Equilibrium moisture content and crystallization of amorphous sucrose and glucose. J. Agr. Food Chem. *4*, 72–77.

MALLOWS, J. H. 1960. Foods of simple structure. Soc. Chem. Ind. Monograph 7, 10–13.

MILLER, R. C., and ZIEGLER, P. T. 1936. How aging affects the distribution of salt in cured hams. Food Inds. *8*, 121–122.

MOHR, W. 1957. Improvement of the consistency and the structure of butter in summer and winter by the technical churning process. Fette Seifen Anstrichmittel *59*, 217–221.

MOYLS, A. W., ATKINSON, F. E., STRACHAN, C. C., and BRITTON, D. D. 1955. Preparation and storage of canned berry and berry-apple pie fillings. Food Technol. 9, 629–634.

MULDER, H. 1953. The consistency of butter. In Foodstuffs: Their Plasticity, Fluidity, and Consistency. Edited by G. W. Scott Blair. Interscience Publishers, New York.

NEVILLE, H. A., EASTON, N. R., and BARTRON, L. R. 1950. The problem of chocolate bloom. Food Technol. 4, 439–441.

NICKERSON, T. A. 1954. Lactose crystallization in ice cream. I. Control of crystal size by seeding. J. Dairy Sci. 37, 1099–1105.

NICKERSON, T. A., and PANGBORN, R. M. 1961. The influence of sugar in ice cream. III. Effect on physical properties. Food Technol. 15, 105–106.

PALMER, K. J., DYE, W. B., and BLACK, D. 1956. X-ray diffractometer and microscopic investigation of crystallization of amorphous sucrose. J. Agr. Food Chem. 4, 77–81.

SALWIN, H. 1959. Defining minimum moisture contents for dehydrated foods. Food Technol. 13, 594–595.

SALWIN, H., and SLAWSON, V. 1959. Moisture transfer in combinations of dehydrated foods. Food Technol. 13, 715–718.

SHIMAZU, F., and STERLING, C. 1961. Dehydration in model systems: Cellulose and calcium pectinate. J. Food Sci. 26, 291–296.

STERLING, C., and SHIMAZU, F. 1961. Cellulose crystallinity and the reconstitution of dehydrated carrots. J. Food Sci. 26, 479–484.

TRESSLER, D. K., and JOSLYN, M. A. 1961. Fruit and Vegetable Juice Processing Technology. Avi Publishing Company, Westport, Conn.

ZOBEL, H. F., and SENTI, F. R. 1959. The bread staling problem. X-ray diffraction studies on breads containing cross-linked starch and a heat-stable amylase. Cereal Chem. 36, 441–451.

CHAPTER 17

Effects of Non-enzymatic Chemical Changes

INTRODUCTION

The role played by non-enzymatic reactions in texture deterioration during storage has not received the attention it deserves. Much more effort has been devoted to the study of the effects of such reactions on flavor (e.g., rancidity), appearance (browning), and nutrients. It does appear that texture is not as subject as flavor and appearance are to change in the absence of native enzymes and micro-organisms. This may be due to the more resistant type of molecule making up the structural elements of natural products as compared with those compounds affecting flavor and color. Most of the textural components are polymers of relative inertness while pigments and flavors are apt to be quite labile.

In this chapter, each type of reaction which has been studied in sufficient detail to permit a worthwhile review to be made will be discussed from the standpoint of its influence on texture.

DENATURATION OF PROTEINS

A definition of denaturation which has received considerable currency is that of Neurath *et al.* (1944), ". . . any non-proteolytic modification of the unique structure of native proteins, giving rise to definite changes in chemical, physical, or biological properties." The predominant agent in these changes may be either physical or chemical. When a protein is denatured, it usually becomes less soluble and loses some of its specific qualities, such as contractility (in muscle fibrils) or catalytic activity (in enzymes). It also exhibits an increased susceptibility to attack by proteolytic enzymes and a higher intrinsic viscosity. Other changes may be observed but they are not often important influences on the texture of foods.

Denaturation is thought to be the expression of configurational changes in the protein molecule. The specific qualities of a protein, such as its immunological properties or its enzymatic activity, result from a highly-ordered spatial relationship of the many different functional groups attached to the skeleton of the polymer or to side chains. This relationship is maintained by hydrogen-bonding and disulfide linkages between groups located in different regions of the molecules as well as by other types of chemical bonds which are not as well understood. Since these bonds are weak, relative to the peptide linkages, for example, changes in the spatial configuration can be induced more easily than changes in the molecular weight or in the functional groups themselves.

Denaturation can be brought about by heat, surface action, radiation, and other physical agents as well as by acids and bases, alkyl sulfates (i.e., some detergents), alcohols, certain hydrophilic ions, etc. Dehydration on a molecular scale usually causes denaturation. Although this chapter is concerned principally with chemical effects, all types of denaturation will be considered.

Heat is the principal denaturant normally affecting the texture of foodstuffs. Most proteins in solution are coagulated at temperatures above 145°F. A protective effect is often exerted by other molecules and ions when the proteins are in complex mixtures, as they are found in nature, but even then temperatures in the range mentioned will coagulate the proteins if the exposure is sufficiently prolonged.

In some instances it appears that denaturation can be partially reversed. There are no clear-cut cases of reversal of denaturation in phenomena predominantly textural, but there seems to have been unequivocal demonstrations of the reversal of heat denaturation of peroxidase in canned peas and similar vegetables. The mechanism is unknown and it is difficult to visualize a process which would re-establish the involved linkage complexus supposedly necessary for the specific configuration of a given protein and for enzymatic activity.

Subsequent to denaturation, the protein molecules tend to polymerize. This association tendency is maximal near the isoelectric point of the protein, at which level the protein forms a coagulum. At other pH's the proteins may remain soluble even though micelles much larger than normal are formed. So far as texture is concerned, denaturation usually results in a shrinkage and a toughening of tissues high in protein. The softening of meats during cooking and the like result mostly from reactions other than denaturation. Inactivation of texture-affecting enzymes during cooking, blanching, and other preparation steps is an important factor in preventing deterioration during the later steps of processing and in storage. Most dehydration methods result in significant denaturation of the proteins even though mild heat is used throughout the processing.

NON-ENZYMATIC BROWNING

The reaction of carbonyl groups from reducing carbohydrates, such as glucose, with amino groups from proteins, peptides, and amino acids results in the formation of colored amorphous polymers of poorly defined structure. Although most of the research activity in this field has been directed toward the elucidation of the flavor and color effects of browning, it appears that detectable effects on texture can occur with moderate degrees of reaction. Such effects might be expected from a consideration of the reactants. Reduction in hydration capacity of the proteins should result from shielding or destruction of the amino groups by the browning reaction. If the reaction is extensive, denaturation might occur with further effects on texture.

Bhatia (1959) found a relationship between browning and mealiness (a texture defect) in dehydrated cooked pork mince. In mince samples which were normal, desugared, or desugared-with-glucose-replaced, development of browning as determined by increase in reflectance, was accompanied by increases in mealiness as determined by sensory tests.

HYDROLYSIS OF STARCHES AND SUGARS

Hydrolysis of sucrose is of little or no importance in texture changes in fruits and vegetables. Dates are an exception, but the reaction in these fruits is catalyzed by an enzyme. In manufactured foods, there are several instances of the dependence of texture on the extent of sucrose hydrolysis. The most striking example occurs in hard candy preparation. Here the hardness, ease of fracture, and resistance to graining is partly a function of the concentration of invert sugar. The invert sugar is usually formed from added sucrose during the cooking period although it may be added as a separate ingredient by some processors.

Sucrose is readily hydrolyzed by acids. In hard candy melts its hydrolysis may be accelerated by adding acid-reacting substances such as potassium hydrogen tartrate or citric acid. Further discussions of these techniques and their effects on the organoleptic properties of hard candies are included in Chapter 9, "Glassy-structured Foods."

Starch, glycogen, and smaller molecular weight polymers of glucose, as well as polymers of fructose such as inulin, are generally rather resistant to non-enzymatic splitting under the conditions usually existing during the processing and storage of foods. In acid foods, retorting or long storage at moderate temperatures may induce some splitting of starch chains. This reaction is rarely effective in altering the texture to a noticeable extent.

Cellulose is inert toward mild chemical reagents and to the physical conditions normally encountered by foodstuffs. The principal storage change which has been observed is the crystallization described in the preceding chapter.

REACTIONS AFFECTING LIGNIN

As discussed elsewhere in this book, Isherwood (1960) has pointed out situations where lignin can, in his opinion, strongly influence the texture of certain fruits and vegetables. Lignin is a relatively reactive substance capable of changes in the absence of enzymatic catalysis and under mild conditions.

Lignin resembles the tannins in that they both contain polyhydroxy aromatic moieties. Many workers regard coniferyl alcohol or some similar compound as the fundamental unit in lignin. Various side chains and functional groups may intervene to change the properties of the polymer. In plants, some of the lignin (perhaps all of it) is loosely bound to the cellulose as an incrustation. In this capacity it is said to give strength and rigidity to the tissues. So far as plant texture is concerned, its action is generally undesirable in that it tends to make the food tougher and more fibrous.

According to Isherwood, lignin will polymerize and combine with a variety of compounds so that it can change the texture of a fruit or vegetable during maturation or storage. Drying of bean pods in the field or subsequently tends to increase lignification.

REACTIONS AFFECTING PECTIN

The gel-forming properties of pectin are influenced by the molecular weight of the molecule, and the extent to which the carboxyl groups are esterified. Depolymerization requires fairly stringent conditions in the absence of appropriate enzymes. The ester linkages are also rather stable under the conditions normally found in foods since they are resistant to hydrolysis in neutral or moderately acidic solutions, although they are readily split by alkalies. Jams and jellies, the principal manufactured foods depending for their textural properties on pectin gels, generally have a pH lying in the range of 2.8 to 3.5. From these considerations it can be deduced that pectin gel-foods (whether manufactured or natural) will not change much in texture as the result of chemical reactions during storage at normal levels of temperature.

The principal texture-affecting changes in these foods are, of course, physical or enzymatic, as discussed in the chapters devoted to such changes. Should a source of cation-contaminant become available, as by the erosion of a metallic container by the acidic aqueous phase of a jelly, it is possible that changes in rheological properties of the food might be observed since gels of this type are notoriously subject to influence by polyvalent metallic ions especially if the methoxy content is relatively low.

OXIDATION

The effect of oxidation on the texture of foodstuffs has been studied very little except in the case of the gluten of wheat flour, where the role of oxidation and reduction in the rheological properties has been extensively investigated. Since these products are raw materials never consumed without further processing, they are beyond the scope of this book. However, the principles elucidated in the investigations may have application to systems which have not been as thoroughly studied. It appears that oxidation by atmospheric oxygen during storage can increase the elasticity and tenacity of the gluten proteins, probably by causing the formation of disulfide linkages between protein molecules at the expense of sulfhydryl groups. If somewhat similar changes occur in gelatin and casein gels, economically important uses could be made of the reaction, but no published information on this point exists.

Oxidation of fats can lead to texture changes. Sterling (1961) attributed the gel-like structure and plastic properties of oxidized cocoa butter to the presence of a significant amount of an amorphous fraction. The difficulty in crystallization ("omega crystallization") is probably due to steric hindrance by the oxidized groups. Changes in the acceptability of fats due to texture deterioration are minor as compared with the effect of flavor changes.

BIBLIOGRAPHY

ANSON, M. L. 1945. Protein denaturation and the properties of protein groups. Advances in Protein Chem. 2, 361–386.

BHATIA, B. S. 1959. Role of products of non-enzymatic browning on development of mealiness in dehydrated cooked pork mince during storage in air. Food Sci., Mysore 8, 309–312.

BRONSON, W. 1951. Technology and utilization of gelatin. Food Technol. 5, 51–54.

CONNELL, J. J. 1957. Some aspects of the texture of dehydrated fish. J. Sci. Food Agr. 8, 526–537.

CRAMER, A. B. 1950. Some problems in the manufacture of hard candy. Food Technol. 4, 400–403.

DOESBURG, J. J. 1957. Relation between the solubilization of pectin and the fate of organic acids during maturation of apples. J. Sci. Food Agr. 8, 206–216.

HAMM, R. 1956. The action of adenosinetriphosphoric acid on hydration and rigidity of post-mortem beef muscle. Biochem. Z. 328, 309–322.

HARVEY, H. G. 1960. Gels, with special reference to pectin gels. Soc. Chem. Ind. Monograph 7, 29–63.

HASHIMOTO, Y., FUKAGAWA, T., NIKI, R., and YASUI, T. 1959. Effect of storage conditions on some of the biochemical properties of meat and on the physical properties of an experimental sausage. Food Research 24, 185–197.

HUNT, S. M. V., and MATHESON, N. A. 1958. The effects of dehydration on actomyosin in fish and beef muscle. Food Technol. 12, 410–416.

ISHERWOOD, F. A. 1955. Texture in fruits and vegetables. Food Manuf. 30, 399–402, 420.

ISHERWOOD, F. A. 1960. Texture of plant tissues. Soc. Chem. Ind. Monograph 7, 135–143.

JOUX, J. L. 1957. Role of pectic substances in the maintenance of firmness in pasteurized apricots. Compt. rend. acad. agr. France 43, 506–513.

LABIE, C. 1958. The use of phosphates in the preservation of meat products. Rec. Med. Vet. Ecole d'Alfort 134, 133–138.

MATTSON, S. 1946. The cookability of yellow peas. A colloid-chemical and biochemical study. Acta Agr. Suecana 2, 185–231.

NEURATH, H., GREENSTEIN, J. P., PUTNAM, F. W., and ERICKSON, J. O. 1944. The chemistry of protein denaturation. Chem. Rev. *34*, 157–265.

NEVILLE, H. A., EASTON, N. R., and BARTRON, L.R. 1950. The problem of chocolate bloom. Food Technol. *4*, 439–441.

OKAMURA, K., MATUDA, T., and YOKOYAMA, M. 1958. Studies on the action of phosphates in "kamaboko" and fish sausage products. I. Various effects of phosphate on qualities of "kamaboko" fish cake. Bull. Japan Soc. Sci. Fisheries *24*, 545–550.

OLLIVER, M. 1950. Factors affecting the jelly grading of pectins. Food Technol. *14*, 370–375.

OLSON, N. F. 1959. Study of the control of the physical structure of pasteurized process cheese spreads. Dissertation Abstr. *19*, 2701–2702.

PERSONIUS, C. J., and SHARP, P. F. 1938. Adhesion of potato tissue cells as influenced by pectic solvents and precipitants. Food Research *4*, 299–309.

PETERS, G. L. 1959. Effect of varying chemical compositions brought about by processing methods on serum viscosity and water retention of tomato purée. Dissertation Abstr. *19*, 3098–3099.

POWERS, J. J., PRATT, D. E., DOWNING, D. L., and POWERS, I. T. 1961. Effect of acid level, calcium salts, monosodium glutamate, and sugar on canned pimientos. Food Technol. *15*, 67–74.

RANKIN, J. C., MEHLTRETTER, C. L., and SENTI, F. R. 1959. Hydroxyethylated cereal flours. Cereal Chem. *36*, 215–227.

SALWIN, H., BLOCH, I., and MITCHELL, J. H., JR. 1953. Dehydrated stabilized egg, importance and determination of pH. Food Technol. *7*, 447–452.

SHERMAN, P. 1961A. The water binding capacity of fresh pork. I. The influence of sodium chloride, pyrophosphate, and polyphosphate on water absorption. Food Technol. *15*, 79–87.

SHERMAN, P. 1961B. The water binding capacity of fresh pork. III. The influence of cooking temperature on the water binding capacity of lean pork. Food Technol. *15*, 90–94.

SMITHIES, R. H. 1960. Effect of chemical constitution on texture of peas. Soc. Chem. Ind. Monograph 7, 119–127.

STERLING, C. 1961. Rheology of cocoa butter. IV. Further studies of "omega" crystallinity. J. Food Sci. *26*, 99–105.

ULRICH, R., and MIMAULT, J. 1956. Transformation of pectic compounds and respiration of pears during ripening. Fruits *11*, 467–470.

CHAPTER 18

Effect of Enzymatic Reactions

INTRODUCTION

In a broad sense, nearly all of the alterations of texture which occur during the maturation of plant foods and food raw materials or in meat (either before or after the slaughter of the animal) could be considered enzymatic, excepting only a few changes such as dehydration. Many texture-influencing transformations happening during processing or spoilage are also enzyme catalyzed. However, this discussion will be concerned only with reactions which have been experimentally related to specific enzymes.

In keeping with the plan of this volume, the effect of processing conditions will not be considered in the present chapter even though they may, in some instances, exert their effect on texture through their influence on enzymatic reactions. The discussion here will be concerned principally with changes occurring during growth, normal storage, and spoilage.

Most of the literature in this field has been concentrated in the areas of meat tenderization or changes in the pectic substances of fruits. A few studies attempting to relate vegetable (especially cucumber) texture to the action of specific enzymes have also been published. On the whole, there does not appear to have been the amount of activity which the subject seems to deserve. Doubtless the lack of interest has been due, at least partially, to the complexity of the systems involved and to the difficulty of quantitating texture changes. A lack of appreciation of the economic importance or fundamental value of such studies has perhaps deterred many potential investigators. As the situation now stands, most of the enzymes involved in texture changes are still unknown or uncharacterized.

MUSCLE TISSUES

The tenderness of meat is dependent upon the state of the contractile substance of the muscle and the amount and distribution of the connective tissue (Paul 1957). The tendering of meat which occurs during storage under conditions designed to retard microbiological attack is a well-known and economically important process. The usual treatment for beef is to hang the meat for as long as four weeks (usually less) in a room maintained at 35°F. To inhibit the activity of molds and bacteria, the room may contain several ultraviolet lamps which reduce the number of micro-organisms on the surface of the meat not only as a result of the direct effect of the radiant energy on micro-organisms but also as a result of the action of the ozone generated from the oxygen of the air. The latter effect is probably the more important under actual conditions since ozone is a very potent germicide and, in addition, can penetrate into regions the ultraviolet rays cannot "see."

Wilson *et al.* (1960A and B) described a method for rapid aging which combines the use of oxytetracycline with high temperatures. Five to ten grams of the antibiotic were given parenterally to the animals 2 to 3 hours before slaughter. The carcasses were stored 24 hours at 110°F. The tenderization which occurred under these conditions was almost as great as that which was obtained when carcasses were aged at 35°F. for two weeks. Panel tests and Warner-Bratzler shear values of steaks removed from the short loin were used for comparison. See Table 26.

TABLE 26

EFFECT OF AGING PROCEDURE ON TEXTURE OF BEEF CARCASSES[1]

Muscle and Aging Procedure	Subjective Tenderness[2]				Shear Test[3]	
	Number of Judgments	Mean	Number of Judgments	Mean	Number of Shears	Mean
Longissimus dorsi						
35°F. for two weeks	118	7.8	118	7.7	77	7.9
110°F. for 24 hours	118	7.5	117	7.5	72	8.5
Semitendinosus						
35°F. for two weeks	122	7.4	118	7.2	97	9.0
110°F. for 24 hours	122	6.9	120	6.9	81	9.7
Semimembranosus						
35°F. for two weeks	126	6.2	126	6.1	109	13.0
110°F. for 24 hours	126	6.5	126	6.3	104	10.8
Biceps femoris						
35°F. for two weeks	109	6.9	109	6.7	118	11.1
110°F. for 24 hours	116	6.5	116	6.4	102	10.8
All muscles						
35°F. for two weeks	475	7.1	471	6.9	401	10.5
110°F. for 24 hours	482	6.8	579	6.8	359	10.1

[1] Adapted from Wilson *et al.* (1960A). Shear values obtained from five carcasses, other values from four carcasses.
[2] A taste panel of eight members rated steaks on a ten-point scale for initial tenderness and for residue.
[3] Warner-Bratzler device was used to determine shear values.

Early interpretations of the cause of these changes postulated a general hydrolysis (autolysis) of proteins with the principal tenderizing effect resulting from a solubilization of collagen. Aging procedures are generally restricted to beef and game. Other muscle tissue foods—pork, shellfish, fish, poultry, etc.—would be subject to flavor deterioration or other deleterious changes when aged in the manner described above, and are usually sufficiently tender when properly cooked.

The changes in protein extractability during post-rigor tenderization of chicken breast muscle were studied by Weinberg and Rose (1960). They suggested that tenderization is not merely random autolysis but results instead from a specific cleavage of an actin association responsible for the maintenance of the muscle matrix. Observations which may bear on this point were reported by Locker (1960) who indicated that the various muscles of the ox go into rigor in widely differing states of contraction as defined by the striation patterns of the myofibrils. The final state of a muscle appears to depend to some extent upon the strain imposed on it in the hung carcass and may be modified by cutting or excising. He found that there was no correlation between the tenderness grading of a muscle and its contraction state in rigor, but attributed this lack of correlation to the dominant effect of the connective tissue. Taste tests on *psoas* muscles which had been cut at death and allowed to shorten showed they were tougher than controls. Locker concluded that relaxed muscles are tenderer than partly contracted muscles and that this effect may be significant in the grading of muscles of low connective tissue content.

Von Hippel *et al.* (1960) reported the results of an enzymatic examination of the structure of the collagen molecule. The kinetics of the collagenase-catalyzed degradation of soluble ichthyocol were followed by pH-stat and colorimetric ninhydrin methods. They indicated that below 81°F. the over-all kinetics can be reduced to the sum of two concurrent reactions, both apparently first order in substrate concentration but differing markedly in rate. Results were interpreted in terms of a rigid, multistranded, interchain hydrogen-bonded structure for the collagen macromolecule in solution. Enzymatic cleavage of single strands leaves the particle relatively intact, but brings about a partial structural collapse by introducing points of increased flexibility.

Development of rigor seems to be accompanied by a loss of the co-enzyme adenosinetriphosphate (ATP). Conversely, this compound has been found to have a tenderizing effect on beef muscles. Hamm (1956) reported that this effect corresponded to an increase in hydration of muscle proteins, while contraction was accompanied by a dehydration. Post-mortem decreases observed in muscle hydration were largely due to a breakdown of ATP, but a lower pH resulting from the accumulation of lactic acid was also partly responsible. From a study of the calcium pyrophosphates, Hamm deduced that ATP, and, to a lesser extent adenosinediphosphate increase the hydration softness of muscle by elimination of the dehydrating calcium ion, with which they form poorly soluble complexes.

Adenosinetriphosphate was also implicated in the development of tenderness in poultry by De Fremery and Pool (1960), who applied several treatments to chicken muscles and determined their effect on toughness as measured by the Warner-Bratzler shear apparatus. Every treatment that resulted in more rapid loss of ATP (i.e., more rapid development of rigor mortis), more rapid drop of pH, and more rapid loss of glycogen also caused increased muscle toughness. Injection of sodium monobromoacetate, which caused rapid loss of ATP but only a small decrease in pH and glycogen, failed to induce toughness. De Fremery and Pool postulated that the relative toughness of cooked muscle in otherwise uniform groups of chickens increases with increasing rate of onset of rigor mortis or with some closely related factor.

Tenderization of tough low-grade cuts of meat by the application of vegetable proteinases has long been practiced. Papain from papaya, bromelin from pineapple, and ficin from figs are some of the enzymes which have been used. Fungal or bacterial sources for the proteinases have also been suggested. Several commercial preparations incorporating papain are now available. It is reasonably certain that these enzymes exert their tenderizing effect through a general solubilization or hydration of the proteins. Some authorities contend that tenderization by these methods results in a texture different than that obtained by "natural" tenderization.

Penny (1960) recommended the use of an accelerated freeze-drying process in conjunction with enzymes for up-grading tough cuts of meat. Freeze-drying yields a porous structure which permits the ready and uniform absorption of proteinase solution. Penny found that dried steaks prepared from cow meat which was tough to the point of inedibility, when reconstituted in tenderizer solution and cooked in a casserole two hours, were found by taste panels to be very acceptable. The most satisfactory enzymes were ficin at a concentration of 0.0025 per cent and papain at 0.005 per cent.

FRUITS AND VEGETABLES

The enzymes acting on pectic substances can have marked effects on the texture of fruits and vegetables, especially the former. Two types of enzymes are especially important. These are pectinesterase, which catalyzes the removal of methoxyl groups from the pectin molecule, and polygalacturonase (pectinase) which hydrolyzes the glycosidic bonds of pectic acid causing a decrease in viscosity and an increase in reducing groups.

Changes in firmness of fruits during maturation have been related to variations in types and amounts of pectic substances by several investigators. Early work by Appleman and Conrad (1926) indicated that softening of peaches during normal ripening paralleled the transformation of protopectin into pectin. Addoms *et al.* (1930) found that maturation of Elberta peaches was accompanied by a thinning and rounding-off of cell walls. In some cases, actual breakdown of the cell walls occurred. These changes were probably due to the enzymatic attack on the protopectin of the cell walls.

The work of Doesburg (1957) is rather typical of more recent studies. He determined the pectin content, degree of esterification, and molecular weight of pectic substances in apples at weekly intervals for some months before and after harvesting. Doesburg concluded that there was no shortening of the chain-length of pectin molecules during solubilization of part of the pectins during ripening. The correlation between changes in solubility of pectin and calcium, the changes in composition of the mixture of organic acids in the fruits, and some evidence of a change in pH of cell walls during this period, were thought to be indications that solubilization of pectin during the ripening of fruits is caused by the movement of calcium in the cell walls.

Huet (1956) showed that decrease of firmness in the pulp of the banana after harvest is associated with a decrease of alcohol insoluble substances, i.e., starch, pectins, cellulose, and hemicelluloses. The decrease of firmness was found to precede the decrease in content of total pectins and protopectins, but to coincide with the decrease in contents of starch and of cellulose plus hemicelluloses. In this case, it would appear that amylolytic and cellulolytic enzymes are more important than pectinolytic enzymes in directing texture changes during ripening.

Jermyn and Isherwood (1956) followed changes occurring during storage of the Conference pear at 77° and 41°F. by extracting the ethanol insoluble cell wall material with cold water and cold alkali. Each fraction was hydrolyzed and chromatographed, and the sugars estimated. Results indicated that the cell wall of the pear is in dynamic equilibrium with the cytoplasm, and polysaccharides are both broken down and synthesized in the course of the physiological changes which occur during the ripening of the fruit.

Joux (1957) stated that the amount of protopectin in apricots is related to the variety and is degraded to pectin as maturation proceeds. Addition of calcium protects protopectin from degradation to some extent. Since apricots contain little pectinesterase and slight amounts of low methoxyl pectin, firmness cannot be improved by addition of calcium to form pectic acid-calcium gels.

Persistence of pectinolytic enzymes through the processing steps required for making frozen orange juice creates storage problems with this product. According to Rouse and Atkins (1955), pectinesterase and pectic substances are important in the production of frozen concentrated juices because of their relation to cloud stability. Results of determinations of these two factors in 221 commercial samples of frozen concentrated orange juice from 27 processors showed a twelve-fold variation in pectinesterase activity and a four-fold variation in amount of pectic substances.

Atkins *et al.* (1953) found that it was necessary to heat orange juice prior to concentration in order to prevent loss of cloud. Bissett *et al.* (1953) related the intensity of heat treatment necessary to obtain cloud stabilization in 1X, 2X, or 4X orange juice to the conditions required to secure substantial inactivation of the pectinesterase. They found that a temperature of 190°F. prevented both loss of cloud and sediment formation. Other pulpy frozen juices, notably tomato, might be expected to show the same reaction.

Ezell and Olsen (1958) found that stabilization temperature had no significant effect on the initial viscosity, cloud or pulp content of concentrates prepared from Valencia oranges. However, after storage at 80°F., viscosity, light transmittance, and amount of centrifuged material differed, depending upon stabilization temperature. As the latter was increased, changes in viscosity and centrifuged material became smaller and the degree of stability increased. These effects are due, of course, to differences in the amount of pectinolytic enzymes inactivated.

Rouse *et al.* (1958) indicated that freeze-damaged oranges and grapefruit decreased in pectinesterase activity and sodium hydroxide-soluble pectin as per cent of total pectin, whereas water soluble pectin, cloud, and viscosity increased. Bissett (1958) also studied the effect of freeze damage on the pectinolytic activity and pectic substances of Parson Brown, Pineapple, and Valencia oranges. He found that varying time intervals between freezing and harvesting did not affect pectinesterase activity, total pectin, soluble pectin, relative viscosity, cloud, stability, or gelation. Juice yields, pectinesterase activity, total pectin and relative viscosity values increased with increasing finisher pressure. It was also found that Valencia concentrates contained less pectinesterase and total pectin, and had lower relative viscosities than did concentrates from Parson Brown and Pineapple varieties.

Pectinolytic enzymes are also of importance in determining cider quality. Cider should be of low viscosity and limpid in appearance. Pollard (1958) stated that, during normal cider production, the pectin enzymes may be derived from contaminants on the fruit, the fruit itself, or the micro-organisms present during fermentation. Pollard's experiments established that pectin breakdown is caused by a combined pectin methylesterase in the fruit itself and a polygalacturonase produced by yeasts. Diluted concentrates or flash pasteurized juices may lack one or both of these enzymes and these ciders may still contain appreciable amounts of pectin.   To secure the desired lower viscosity and absence of cloud, addition of pectic enzymes is recommended.

Hoogzand and Doesburg (1961) found that the degree of esterification of pectic substances in cauliflower is decreased by activation of pectin-esterase during low temperature-long time blanching.   As compared with high temperature-short time blanching, the low temperature-long time blanch causes greater firmness, and this effect is enhanced by additions of calcium salts and acid.   Reduction of pH from the normal 6.0 to 5.0 after blanching and before canning did not improve firmness.

Adventitious polygalacturonase has proved to be the cause of texture deterioration in two widely different brined products, cherries and cucumbers.   Brined   cherries   are   the   raw   products   from   which maraschino-type cherries are prepared.   In contrast to most other processed foods, the main consideration in brined cherries is the texture or firmness of the fruit, the color and flavor being added artificially during the maraschino manufacturing process (Bullis and Wiegand 1931).   The desirable firmness of texture is primarily dependent upon the natural pectin present (Kertesz 1951).   When the pectin of brined cherries is destroyed, they become soft and unsuitable for the preparation of mara-schino cherries (McCready and McComb 1954).   The destruction of texture which occasionally occurs in commercial lots is undoubtedly due to the action of a polygalacturonase, such as that described by Phaff and Demain (1956).   The source of this enzyme is still not known but micro-organisms and/or plant diseases are suspected (Ross et al. 1958).

Nacconol (a commercial preparation of sodium alkyl aryl sulfonate) was found by Steele and Yang (1960) to be an effective inhibitor of poly-galacturonase in model systems and was then successfully applied to retarding the activity of this enzyme in brined cherries (Yang et al. 1960). The latter investigators found that it was not feasible to inhibit polyga-lacturonase by lowering the brine pH to the level (pH 1.6 or lower) at which the enzyme is inactive.   The brine is usually made up of one and one-fourth per cent sulfur dioxide (a preservative and bleaching agent) and three-fourths of one per cent calcium hydroxide or calcium chloride. The desired brine pH can be obtained by using calcium chloride, but such brines produce cherries with undesirably tough texture and low crispness, and cherries at low pH also tend to crack and the skin is easily injured.   Therefore, calcium hydroxide is most commonly used even though the pH 3.0 brines (approximately) so obtained permit poly-galacturonase to function.   However, Yang et al. successfully inactivated the enzyme by adding 0.01 to 0.025 per cent commercial alkyl aryl sul-fonate to the brine.   Only trace amounts of the inhibitor remain in the finished product.

Cucumbers normally have a crisp texture when picked. Ideally, crispness is retained substantially undiminished during brine fermentation and storage. In some cases however, this texture characteristic is lost during processing with a consequent reduction in the acceptability of the pickles. Bell, Etchells, and their co-workers at the Food Fermentation Laboratory of the U. S. Department of Agriculture have done outstanding work in identifying the cause of this condition and specifying remedies for it. They (Bell *et al.* 1950) attributed softening in salt-stock to the action of polygalacturonase but indicated that a cellulolytic enzyme was frequently present in faulty lots. Bell *et al.* (1955) described methods for differentiating and estimating the two types of enzymes.

The bulk of the evidence at the present time (see Bell and Etchells reference, also Demain and Phaff 1957) tends to implicate both pectinolytic and cellulolytic enzymes in softening and to point to the cucumber flowers as the major (or only significant) source of these enzymes. The deteriorative biocatalysts are not thought to be native to the flowers but to arise from the activities of fungi which grow on the flowers prior to harvest.

Hamilton and Johnston (1961A and B) isolated 99 gram-positive sporeforming bacteria cultures and 15 yeast cultures from commercial cucumber salt-stock brine and found that none produced pectolytic enzymes under simulated commercial conditions. However, isolates belonging to ten genera of filamentous fungi were found to be moderately to highly pectolytic when tested under the same conditions. None of the organisms produced any pectinesterase in pure culture study. These findings would seem to implicate the cucumber as the source of pectinesterase.

At least four methods have been suggested (Bell *et al.* 1955) for eliminating or reducing the concentration of softening enzymes in commercial cucumber brines. These are: (1) mechanical removal of flowers from the cucumbers before brining; (2) development of new cucumber varieties with a minimum of retained flowers; (3) development of draining procedures to reduce enzyme content of brines before the enzymes can affect cucumber firmness; and (4) inactivation of the enzymes in brine with specific, non-toxic inhibitors.

Grape leaves had been recommended for many years by certain authors as a desirable additive for home-pickled cucumbers although their function was somewhat obscure. However, Etchells *et al.* (1958) showed that a crude extract of Scuppernong grape (*Vitus rotundifolia*) leaves effectively reduced the activity of pectinolytic and cellulolytic enzymes extracted from cucumber flowers. The reduction of activity of the two enzymes was directly related to the inhibitor concentration, and higher

levels of inhibitor resulted in an increase in the firmness of the fermented
cucumbers (salt-stock). See Fig. 29.

Bell *et al.* (1960) further characterized a cellulase inhibitor from the
mature leaves of six varieties of Muscadine grapes. Extracts containing
the inhibitor were stable to heat, to weak acid and alkali, and to such
protein-precipitating agents as trichloroacetic acid. The inhibitor was
non-dialyzable through cellophane or collodion membranes against water
or weakly buffered solutions. They concluded that the cellulase-inhibit-
ing substance is not related in structure to carbohydrates or proteins and
appears to be a high-molecular weight organic compound. Later work
seemed to indicate that the inhibitor was similar to the tannins in chemical
composition.

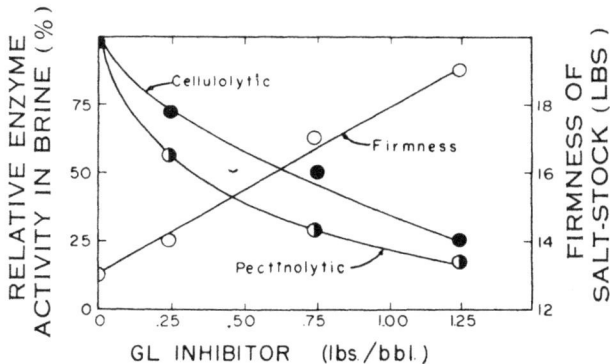

Fig. 29. Effect of Grape-Leaf Inhibitor on Maximum
Pectinolytic and Cellulolytic Enzyme Activity in Cu-
cumber Brines as Related to Firmness of Cured Salt-
Stock

Other Enzymes

Mattson *et al.* (1950) found a decided increase in the cookability of
yellow peas with an increase in the humidity of the air and in the moisture
content of the peas after storage for 3, 6, or 12 months. They also
observed an increased orthophosphate content after storage and concluded that phytin was broken down by phytase activity and the resultant redistribution of calcium caused the texture changes.

Bode (1961) described a commercial process for improving (tenderizing) canned pea texture through the addition of amylases to the brine
before cooking. This investigator used an enzyme which remained partially active up to 212°F. and therefore exerted some effect during the
early part of the retorting. This thermostability suggests that the enzyme

was of bacterial origin although it is not so described in the article. The enzyme acts by breaking down the highly-polymerized starch which is partly responsible for toughness in peas. An improvement in flavor also results, so it is said, from the formation of dextrins and maltose.

Evidently, invertases are not responsible for significant texture changes in fruits and vegetables, or, if so, the relationship has not been established as yet. However, Cook and Furr (1953) showed that the kinds and relative amounts of sugar in cured dates affected the texture of this fruit. The high sucrose dates examined by Cook and Furr were usually firm, while those with relatively large amounts of reducing sugars tended to be soft. Semi-dry varieties were intermediate in texture. The differences in sugar composition are due, in part, at least, to the extent to which sucrose is enzymatically hydrolyzed during the curing process.

CEREALS

So far as their effect on texture is concerned, amylolytic enzymes can be regarded as the cereal counterparts of the pectinolytic and cellulolytic enzymes of fruits and vegetables and the proteinases of meats. In the relatively dry state in which most cereals are usually found, amylases are not active. When the cereal is brought to a higher moisture content, as in the preparation of dough or similar processes, the enzyme can resume its activity, but the baking or boiling (as in the case of alimentary pastes) which succeeds doughing-up again inactivates the enzyme. This series of events means that amylases have a favorable environment for too short a time to cause appreciable texture changes, in most cases.

From time to time suggestions are made that addition of heat-stable amylases to bread doughs would be a good method for retarding the staling of the finished product. Bread staling is considered to be due primarily to retrogradation of the starch, and the molecule-splitting which results from amylase activity might interfere with the process. Blinc (1956) compared the staling rates of bread loaves made with and without four per cent beta-amylase and stored at 86°F. and 70 to 90 per cent relative humidity. The bread containing amylase showed improved compressibility and softness of the crumb.

Several years ago some preliminary attempts were made by personnel at the Quartermaster Food and Container Institute to use heat-stable amylase (from bacterial sources) in canned bread for the Armed Forces. This product always reaches its ultimate degree of texture staling before the consumer receives it. It was postulated by these workers that the presence of a small amount of amylase, which had survived the temperatures reached during the baking process, would be able (over a period of weeks or months) to offset at least partially the hardening or toughening effect of starch retrogradation. Only limited success was achieved in this project and it was finally dropped. However, the idea still seems attractive.

SUMMARY

Proteinases have received the most study as texture-influencing enzymes in meat. Their principal effect seems to be on the connective tissue which causes much of the unpleasant textural quality of low grade meats. Enzymes which act directly on the contractile fibriles, perhaps in conjunction with the co-enzymes ATP and ADP, are doubtless active in many instances but have not been as well characterized as the proteinases.

Pectinolytic and cellulolytic enzymes are important in determining the texture of fruits and vegtables. The degradation of protopectin with a consequent softening of the tissues is apparently a common reaction in maturing fruits. Pectinesterases and polygalacturonases are effective both in maturation and spoilage. Enzymes affecting cellulose may be active in many plant foods, but they have not been as thoroughly studied as the pectinolytic enzymes. Phytases are said to be important in influencing the cookability and tenderness of certain dried legumes.

Amylases are the typical texture-affecting enzymes of cereals. The staling reactions in baked products can be retarded by such biocatalysts. They are doubtless important in alimentary paste manufacture although little study has been made of this problem. Native proteinases are present in very low levels in cereals and probably have no perceptible effect on texture, but additions of fungal enzymes of this type have been suggested and used in bread doughs.

BIBLIOGRAPHY

Addoms, R. M., Nightingale, G. T., and Blake, M. A. 1930. Development and ripening of peaches as correlated with physical characteristics, chemical composition, and histological structure of the fruit flesh. II. Histology and microchemistry. New Jersey Agr. Expt. Sta. Bull. No. 507.

Appleman, C. O., and Conrad, C. M. 1926. Pectic constituents of peaches and their relation to softening of the fruit. Maryland Univ. Agr. Expt. Sta. Bull. 238.

Atkins, C. D., Rouse, A. H., Huggart, R. L., Moore, E. L., and Wenzel, F. W. 1953. Gelation and clarification in concentrated citrus juices. III. Effect of heat treatment of Valencia orange and Duncan grapefruit juices prior to concentration. Food Technol. 7, 62–66.

Bartholomew, E. T., and Sinclair, W. B. 1941. Unequal distribution of soluble solids in the pulp of citrus fruits. Plant Physiol. 16, 293–306.

Bartholomew, E. T., Sinclair, W. B., and Turrell, F. M. 1941. Granulation of Valencia oranges. Calif. Univ. Agr. Expt. Sta. Bull. 647.

Bell, T. A., Aurand, A. W., and Etchells, J. L. 1960. Cellulase inhibitor in grape leaves. Botan. Gaz. 122, 143–148.

BELL, T. A., and ETCHELLS, J. L. 1960. Influence of salt on pectinolytic softening of cucumbers. Food Research 25, 84–90.

BELL, T. A., ETCHELLS, J. L., and JONES, I. D. 1950. Softening of commercial cucumber salt-stock in relation to polygalacturonase activity. Food Technol. 4, 157–163.

BELL, T. A., ETCHELLS, J. L., and JONES, I. D. 1955. A method for testing cucumber salt-stock brine for softening activity. U. S. Dept. Agr. ARS 72–5.

BISSETT, O. W. 1958. Processing freeze-damaged oranges. Proc. Florida State Hort. Soc. 71, 254–259.

BISSETT, O. W., VELDHUIS, M. K., and RUSHING, N. B. 1953. Effect of heat treatment temperature on the storage life of Valencia orange concentrates. Food Technol. 7, 258–260.

BLINC, M. 1956. Some observations concerning the effect of beta-amylase on the staling of bread. Brot u. Gebäck 10, 249–252

BODE, H. E. 1961. Enzymes act as tenderizers. In Practical New Canning and Freezing Methods. Chilton Publications, Philadelphia.

BRISKEY, E. J., BRAY, R. W., HOEKSTRA, W. G., PHILLIPS, P. H., and GRUMMER, R. H. 1960. Effect of high protein, high fat, and high sucrose rations on the water-binding and associated properties of pork muscle. J. Animal Sci. 19, 404–411.

BUCH, M. L., SATORI, K. G., and HILLS, C. H. 1961. The effect of bruising and aging on the texture and pectic constituents of canned tart cherries. Food Technol. 15, 526–531.

BULLIS, D. E., and WIEGAND, E. H. 1931. Bleaching and dyeing Royal Anne cherries for maraschino or fruit salad use. Oregon State Coll. Agr. Expt. Sta. Bull. 275.

COOK, J. A., and FURR, J. R. 1953. Kinds and relative amounts of sugar and their relation to texture in some American-grown date varieties. Proc. Am. Soc. Hort. Sci. 61, 286–292.

DE FREMERY, D., and POOL, M. F. 1960. Biochemistry of chicken muscle as related to rigor mortis and tenderization. Food Research 25, 73–87.

DEMAIN, A. L., and PHAFF, H. J. 1957. Cucumber curing: softening of cucumbers during curing. J. Agr. Food Chem. 5, 60–63.

DODGE, J. W., and STADELMAN, W. J. 1959. Post mortem aging of poultry meat and its effect on the tenderness of the breast muscles. Food Technol. 13, 81–84.

DOESBURG, J. J. 1957. Relation between the solubilization of pectin and the fate of organic acids during maturation of apples. J. Sci. Food Agr. 8, 206–216.

ETCHELLS, J. L., BELL, T. A., and WILLIAMS, C. F. 1958. Inhibition of pectinolytic and cellulolytic enzymes in cucumber fermentations by Scuppernong grape leaves. Food Technol. 12, 204–208.

EWELL, A. W. 1940. The tendering of beef. Refrig. Eng. 39, 237–240.

EZELL, G. H., and OLSEN, R. W. 1958. Effect of stabilization temperature on the viscosity and stability of concentrated orange juices. Proc Florida State Hort. Soc. 71, 186–189.

FRANCIS, F. J., AMLA, B. L., and KIRATSOUS, A. 1959. Control of exudation in pre-peeled French-fry potatoes with antibiotics. Food Technol. 13, 485–488.

HALLER, M. H., HARDING, P. L., and ROSE, D. H. 1933. The interrelation of firmness, dry weight, and respiration in strawberries. Proc. Am. Soc. Hort. Sci. 29. 330–334.

HAMILTON, I. R., and JOHNSTON, R. A. 1961A. Studies of cucumber softening under commercial salt-stock conditions in Ontario. I. Incidence and pattern of activity of pectolytic enzymes. Appl. Microbiol. 9, 121–127.

HAMILTON, I. R., and JOHNSTON, R. A. 1961B. Studies of cucumber softening under commercial salt-stock conditions in Ontario. II. Pectolytic micro-organisms isolated. Appl. Microbiol. 9, 128–133.

HAMM, R. 1956. The action of adenosinetriphosphoric acid on hydration and rigidity of post-mortem beef muscle. Biochem. Z. 328, 309–322.

HEINZE, P. H., KIRKPATRICK, M. E., and DOCHTERMAN, E. F. 1955. Cooking quality and compositional factors of different varieties from several commercial locations. U. S. Dept. Agr. Tech. Bull. 1106.

HOOGZAND, C., and DOESBURG, J. J. 1961. Effect of blanching on texture and pectin of canned cauliflower. Food Technol. 15, 160–163.

HUET, R. 1956. Note on the biochemical significance of firmness in the case of the pulp of the banana. Fruits 11, 395–399.

JACOBS, M. B. 1959. The Chemistry and Technology of Food and Food Products. Interscience Publishers, New York.

JERMYN, M. A., and ISHERWOOD, F. A. 1956. Changes in the cell-wall of the pear during ripening. Biochem. J. 64, 123–132.

JOUX, J. L. 1957. Role of pectic substances in the maintenance of firmness in pasteurized apricots. Compt. rend. acad. agr. France 43, 506–513.

KERTESZ, Z. I. 1951. The Pectic Substances. Interscience Publishers, New York.

KIEFER, F. 1961. A new oxidative mechanism in the deteriorative changes of orange juice. Food Technol. 15, 302–305.

LOCKER, R. H. 1960. Degree of muscular contraction as a factor in tenderness of beef. Food Research 25, 304–307.

MATTSON, S., ÅKERBERG, E., ERICKSSON, E., KOUTLER-ANDERSSON, E., and VATHTRAS, K. 1950. Factors determining the composition and cookability of peas. Acta Agr. Scand. 1, 40–61.

McCREADY, R. M., and McCOMB, E. A. 1954. Texture changes in brined cherries. Western Canner Packer 46, No. 12, 17–19.

NICHOLAS, R. C., and PFLUG, I. J. 1960. Effects of high temperature storage on the quality of fresh cucumber pickles. Glass Packer 39, No. 4, 35, 38–39, 65.

PANGBORN, R. M., VAUGHN, R. H., YORK, G. K., III, and ESTELLE, M. 1959. Effect of sugar, storage time, and temperature on dill pickle quality. Food Technol. 13, 489–492.

PAUL, P. C. 1957. Tenderness of beef. J. Am. Dietetic Assoc. 33, 890–894.

PENNY, I. F. 1960. Up-grading of low-grade meat. Chem. Ind. 1960. 288–289.

PHAFF, H. J., and DEMAIN, A. L. 1956. The unienzymatic nature of yeast poly-galacturonase. J. Biol. Chem. 218, 875–887.

POLLARD, A. 1958. Fermentation of diluted concentrates. In Intern. Fed. Fruit Juice Producers' Symp. (Bristol) 1958, 351–360.

ROSS, E., YANG, H., and BREKKE, J. E. 1958. Preliminary report on the brined cherry project, 1958 season. Wash. State Coll. Agr. Expt. Stas. Circ. 340.

ROUSE, A. H., and ATKINS, C. D. 1955. Pectin esterase and pectin in commercial citrus juices as determined by methods used at the Citrus Experiment Station. Univ. Florida Agr. Expt. Sta. Bull. 570.

ROUSE, A. H., ATKINS, C. D., and MOORE, E. L. 1958. Chemical characteristics of citrus juices from freeze-damaged fruit. Proc. Florida State Hort. Soc. 71, 216–219.

SINCLAIR, W. B., and JOLLIFFE, W. A. 1961. Chemical changes in the juice vesicles of granulated Valencia oranges. J. Food Sci. 26, 276–282.

SON, C. H. 1960. Microscopical study of structural changes of peaches and pears during softening. Dissertation Abstr. 20, 1316.

STEELE, W. F., and YANG, H. Y. 1960. The softening of brined cherries by poly-galacturonase, and the inhibition of polygalacturonase in model systems by alkyl aryl sulfonates. Food Technol. 14, 121–126.

STIER, E. F., BALL, C. O., and MACLINN, W. A. 1956. Changes in pectic substances of tomatoes during storage. Food Technol. 10, 40–43.

VON HIPPEL, P. H., GALLOP, P. M., SEIFTER, S., and CUNNINGHAM, R. S. 1960. An enzymatic examination of the structure of the collagen molecule. J. Am. Chem. Soc. 82, 2774–2786.

WANG, H., WEIR, C. E., BIRKNER, M. L., and GINGER, B. 1958. Studies on enzymatic tenderization of meat. III. Histological and panel analyses of enzyme preparations from three distinct sources. Food Research 23, 423–438.

WEINBERG, B., and ROSE, D. 1960. Changes in protein extractibility during post-rigor tenderization of chicken breast muscle. Food Technol. *14*, 376–378.

WEIR, C. E., WANG, H., BIRKNER, M. L., PARSONS, J., and GINGER, B. 1958. Studies on enzymatic tenderization of meat. II. Panel and histological analyses of meat treated with liquid tenderizers containing papain. Food Research *23*, 411–422.

WILSON, G. D., BROWN, P. D., POHL, C., WEIR, C. E., and CHESBRO, W. R. 1960A. A method for the rapid tenderization of beef carcasses. Food Technol. *14*, 186–189.

WILSON, G. D., BROWN, P. D., WEIR, C. E., POHL, C. V., CHESBRO, W. R., and GINGER, B. 1960B. Studies on high temperature aging of beef. American Meat Institute Foundation Bull. *44*.

WOODMANSEE, C. W., McCLENDON, J. H., and SOMERS, G. F. 1959. Chemical changes associated with the ripening of apples and tomatoes. Food Research *24*, 503–514.

YANG, H. Y., STEELE, W. F., and GRAHAM, D. J. 1960. Inhibition of polygalacturonase in brined cherries. Food Technol. *14*, 644–647.

Index

283

www.ingramcontent.com/pod-product-compliance
Lightning Source LLC
Chambersburg PA
CBHW031945080426
42735CB00007B/263